Data Reduction
and
Error Analysis
for the
Physical Sciences

Philip R. Bevington

Associate Professor of Physics · Case Western Reserve University

DATA REDUCTION
AND ERROR ANALYSIS
FOR THE
PHYSICAL SCIENCES

PHILIP R. BEVINGTON
Associate Professor of Physics
Case Western Reserve University

DATA REDUCTION
AND ERROR ANALYSIS
FOR THE
PHYSICAL SCIENCES

McGRAW-HILL BOOK COMPANY
New York   San Francisco   St. Louis
Toronto   London   Sydney

**DATA REDUCTION AND ERROR ANALYSIS
FOR THE PHYSICAL SCIENCES**

Printed in the United States of America.

Library of Congress catalog card number: 69–16942

24 25 26 27 BKP BKP 8 9

TO JOAN

PREFACE

The purpose of this text is to provide an introduction to the techniques of data reduction and error analysis commonly employed by individuals doing research in the physical sciences and to present them in sufficient detail and breadth to make them useful for students throughout their undergraduate and graduate studies. The presentation is developed from a practical point of view, including enough derivation to justify the results, but emphasizing the methods more than the theory.

The level of primary concern is that of junior and senior undergraduate laboratory where a thorough study of these techniques is most appropriate. The treatment is intended to be comprehensive enough to be suitable for use by graduate students in experimental research who would benefit from the generalized methods for linear and non-linear least-squares fitting and from the summaries of definitions and techniques.

At the same time, the introduction to the material is made self-supporting in that no prior knowledge of the methods of statistical evaluation is assumed; the material of each section is developed from first principles. A discussion of differential calculus and manipulation of matrices and determinants is included in the appendixes to supplement their use in the text.

The emphasis, however, is toward the application of more general techniques than are usually presented in undergraduate laboratories. With the proliferation of computers and their use in research laboratories, it is important that sophisticated concepts of data reduction be introduced. Computer routines written explicitly for each section are used throughout to illustrate in a practical way both the concepts and the procedures discussed.

The first five chapters introduce the concepts of errors, uncertainties, probability distributions, and methods of optimizing the estimates of parameters characterizing observations of a single variable. Chapters 6 to 9 deal with the problem of fitting, analytically, complex functions to observations of more than one variable, including estimates of the resulting uncertainties and tests for optimizing the functional form of the fit. The last third of the text contains a description of techniques for searching for the best fit to data with arbitrary functions. Techniques for manipulating data or extracting information without fitting are also discussed.

Computer programs The primary purpose of the computer routines is to clarify the presentation, but they are meant to be usable for calculations as well. They are written as subroutines and function subprograms to provide flexibility for the user in applying them to his own data. The format of the routines is similar to that of the IBM Scientific Subroutine Package for the IBM 360 computer system, including considerable commentary to define the parameters used and to describe the flow of the routine.

The routines are written in Fortran IV, but they are compatible with Fortran II except for the use of double precision and missing suffixes in the names of library functions, which are noted in the program descriptions. The sizes of most of the arrays are to be specified by the user and they are dimensioned in the routines with a size of 1. For most versions of Fortran, the dimension size in a subprogram is a dummy argument for one-dimensional arrays and need not correspond to the actual size used. Two-dimensional arrays are always dimensioned explicitly; these arrays and those which are wholly contained within the routines are assigned dimensions such that the routines can handle up to 10 terms of a fitting function and up to 100 data points. These dimensions may be increased for larger input dimensions or decreased for more efficient use of memory. All input and output variables are specified as arguments of the calling statement.

These routines have been debugged in both Fortran IV and Fortran II versions on small and large computers. They are intended to be usable operating routines and are reasonably efficient. Their most important function, however, is to serve as a framework on which to build and modify routines to serve the specific needs of the user.

Acknowledgments   I am deeply indebted to members of the Stanford University Nuclear Physics Group and especially to Drs. A. S. Anderson, T. R. Fisher, D. W. Heikkinen, R. E. Pixley, F. Riess, and G. D. Sprouse for many helpful comments, suggestions, and techniques. The use of the SCANS (Stanford Computers for the Analysis of Nuclear Structure) PDP-7 and PDP-9 computers for the development of computer routines and the assistance of Wylbur Wryght of the Stanford Computation Center are gratefully acknowledged. I have benefitted from comments by Prof. Jay Orear and especially from many invaluable suggestions and criticisms of Prof. Hugh D. Young.

I wish also to express my appreciation to Mrs. Pat Johnson for her care and diligence in typing and to Mrs. H. Mae Sprouse for her artistic work on the illustrations. Special appreciation is due to my wife, Joan, and our children, Ann and Mark, for enduring so patiently the division of my attention during the writing of this book.

Philip R. Bevington

TABLE OF CONTENTS

APPENDIX C

GRAPHS AND TABLES

CHAPTER ONE

SYSTEMATIC AND
RANDOM ERRORS

1-1 ERRORS

It is a well-established rule of scientific investigation that the first time an experiment is performed the results bear all too little resemblance to the "truth" being sought. As the experiment is repeated, with successive refinements of technique and method, the results gradually and asymptotically approach what we may accept with some confidence to be a reliable description of events. Some investigators have gone so far as to assert that nature is loath to give up her secrets without a considerable expenditure of effort on our part, and that it is a fundamental fact of life that first steps in experimentation are bound to fail. Whatever the reason, it is certainly true that for all physical experiments, errors and uncertainties exist which must be reduced by improvements in techniques and ideas and then estimated to establish the validity of results.

Error is defined by Webster as "the difference between a calculated or observed value and the true value." Usually, of course, we do not know what the "true" value is or there would be no reason for performing the experiment. But we often do know approximately what it should be, either from earlier experimentation along the same line or from other theoretical or experimental approaches. Such approximations can yield an indication of whether our result is of the right order of magnitude, but we also need some systematic way to determine from the data themselves how much confidence we can have in our experimental results.

There is one class of errors which we can deal with immediately: that which originates from mistakes or blunders in computation or measurement. Fortunately, these sources of error are usually apparent either as obviously incorrect data points or as results which are not reasonably close to the expected results. They are classified as *illegitimate errors* which can be corrected by performing the erroneous operations again correctly.

Systematic errors   There is another class of errors which is not so easy to detect and for which statistical analysis is not generally useful. This is the class of *systematic errors*, such as those which result reproducibly from faulty calibration of equipment or from bias on the part of the observer. These errors must be estimated from an analysis of the experimental conditions and techniques. In some cases, corrections can be made to the data to compensate for systematic errors where the type and extent of the error is known. In other cases, the uncertainties resulting from these errors must be estimated and combined with uncertainties from statistical fluctuations.

EXAMPLE 1-1   A student measures a table top with a steel meter stick and finds that the average of his measurements yields a result of 1.982 m for the length of the table. He subsequently learns that the meter stick was calibrated at 25°C and has an expansion coefficient of .0005/°C. Since his measurements were made at a temperature of 20°C, he multiplies his results by $1 - 5(.0005) = 0.9975$ so that his new determination of the length is 1.977 m.

When the same student repeats the experiment, he discovers that his technique for reading the meter stick was faulty in that he did not always read the divisions from directly above. By experimentation he determines that this consistently results in a reading which is 2 mm too short. With this correction, his final result is 1.979 m.

Accuracy vs. precision   There is considerable confusion among students as to the meaning of and difference between the terms accuracy and precision. To add to this confusion, Webster defines them equally. In scientific investigation, however, they are assigned distinctly different meanings which must be kept separate.

The *accuracy* of an experiment is a measure of how close the result of the experiment comes to the true value. Therefore, it is a measure of the correctness of the result. The *precision* of an experiment is a measure of how exactly the result is determined, without reference to what that result means. It is also a measure of how reproducible the result is. The *absolute precision* indicates the magnitude of the uncertainty in the result in the same units as the result. The *relative precision* indicates the uncertainty in terms of a fraction of the value of the result.

For example, in the experiment of Example 1-1, the first result was given with a fairly high precision. The table top was found to be 1.982 m long, indicating an absolute precision on the order of 1 mm and a relative precision on the order of $\frac{1}{2000}$. The corrections to this result were meant to improve the accuracy by compensating for known deviations of the first result from the best estimate possible. These corrections did not improve the precision at all, but did, in fact, worsen the estimated precision because the corrections were themselves only estimates of the exact corrections.

It is obvious that we must consider accuracy and precision

simultaneously for any experiment. It would be a waste of time and energy to determine a result with a high precision if we knew the result would be highly inaccurate. Conversely, a result cannot be considered to be extremely accurate if the precision is low. For example, if the length of the table is quoted as 2. m, the answer is undoubtedly accurate but the amount of information available is limited by the fact that such a precision only specifies the length to be between 1.5 and 2.5 m long. Similarly, if the length of the table is known to be exactly 2.000 m, there would be no point in improving the experimental precision so long as the inaccuracy remains about 20 mm.

Significant figures and round-off The precision of an experimental result is implied by the way in which the result is written, though it should generally be quoted specifically as well. To indicate the precision, we write a number with as many digits as are significant. The number of *significant figures* in a result is defined as follows:

1. The leftmost nonzero digit is the most significant digit.
2. If there is no decimal point, the rightmost nonzero digit is the least significant digit.
3. If there is a decimal point, the rightmost digit is the least significant digit, even if it is a 0.
4. All digits between the least and most significant digits are counted as significant digits.

For example, the following numbers each have four significant digits: 1,234; 123,400; 123.4; 1,001, 1,000., 10.10, 0.0001010, 100.0. If there is no decimal point, there are ambiguities when the rightmost digit is a 0. For example, the number 1,010 is considered to have only three significant digits even though the last digit might be physically significant. To avoid this ambiguity, it is better to supply decimal points or write such numbers in exponent form as an argument in decimal notation times the appropriate power of 10. Thus, our example of 1,010 would be written as 1,010. or 1.010×10^3 if all four digits are significant.

When quoting results of an experiment, the number of significant figures given should be approximately one more than that dictated by the experimental precision. The reason for including the extra digit is that in computation one significant figure is sometimes lost. Errors introduced by insufficient precision in calculations are classified as illegitimate. If an extra digit is specified for all numbers used in the computation, the original precision will be retained to a greater extent. For example, in the experiment of Example 1-1, if the absolute precision of the result is 10 mm, the third figure is known with an uncertainty of ± 1 and the fourth figure is not really known at all. We would be barely justified in specifying four figures for computation. If the precision is 2 mm, the third digit is known quite well and the fourth figure is known approximately. We are justified in quoting four figures, but probably not justified in quoting five figures since we cannot even have much confidence in the value of the fourth figure.

When insignificant digits are dropped from a number, the last digit retained should be rounded off for the best accuracy. To round off a number to a smaller number of significant digits than are specified originally, truncate the number to the desired number of significant digits and treat the excess digits as a decimal fraction. Then

1. If the fraction is greater than $\frac{1}{2}$, increment the least significant digit.
2. If the fraction is less than $\frac{1}{2}$, do not increment.
3. If the fraction equals $\frac{1}{2}$, increment the least significant digit only if it is odd.

In this manner, the value of the final result is always within half the least significant digit of the original number. The reason for rule (3) is that in many cases the fraction equals either 0 or $\frac{1}{2}$ and consistently incrementing the least significant digit for a fraction of $\frac{1}{2}$ would lead to a systematic error. For example, 1.235 and 1.245 both become 1.24 when rounded off to three significant figures, but 1.2451 becomes 1.25.

Random errors The *accuracy* of an experiment, as we have defined it, is generally dependent on how well we can control or compensate for *systematic* errors. These are the errors which will make our results different from the "true" values with reproducible discrepancies. The *precision* of an experiment is dependent on how well we can overcome or analyze *random* errors. These are the fluctuations in observations which yield results that differ from experiment to experiment and that require repeated experimentation to yield precise results. A given accuracy implies a precision at least as good, and, therefore, depends to some extent on random errors also.

The problem of reducing random errors is essentially one of improving the experiment and refining the techniques as well as simply repeating the experiment. If the random errors result from instrumental uncertainties, they can be reduced by using more reliable and more precise measuring instruments. If the random errors result from statistical fluctuations of counting finite numbers of events, they can be reduced by counting more events.

In general, however, we will be interested in extracting the maximum amount of useful information from the data on hand without benefit of being able to repeat the experiment with more or better equipment. We will be concerned, therefore, with the problem of extracting from the data the most reasonable estimates of theoretical parameters and of estimating what the random errors are, what their effects on the results are, and what confidence we can have in the final results.

1-2 UNCERTAINTIES

The term *error* signifies a deviation of the result from some "true" value. Often we cannot know what that true value is, and we can consider only *estimates* of the errors inherent in the experiment. If we repeat the experiment, the results may well differ from those of the first attempt. We can express this difference as a *discrepancy* between the two results. The fact that a discrepancy arises is due to the fact that we can determine our results only with a given *uncertainty*. Throughout this book we will ignore the problem of systematic errors and concentrate on the random errors resulting from uncertainties in our results.

There are two classes of uncertainties. The first and most prominent type are those which result from fluctuations in repeated measurements of data from which the results are calculated. The other class results from the fact that we may not always know what the appropriate theoretical formula is for expressing the result. For example, if we measure the *length* of a table, we know that the result must be a number with the units of length. Any uncertainties, aside from systematic errors, come from the fact that our observations fluctuate from trial to trial. With an infinite number of measurements we might be able to estimate the length very precisely, but with a finite number of trials there will be a finite uncertainty.

If we were to describe the *shape* of an oval table, however, we would be faced with uncertainties both in the measurement of position of the edge of the table at various points and in the form of the equation used to describe the shape, whether it be circular, elliptical or whatever. Therefore, we will be concerned in the chapters to come with a comparison of the actual distribution of the measured data points with that which we would expect on the basis of a theoretical model. This comparison will help to indicate whether our method of extracting the results is valid or needs modification.

Probable error We introduce the term *probable error* to indicate the magnitude of the error which we estimate we have made in our determination of the results. This does not mean that we expect our results to be wrong by this amount. It means, instead, that if our answer is wrong, it *probably* won't be wrong by more than the probable error. The magnitude of the probability, however, is 50%, which means that the difference between the answer and the "true" value probably isn't less than the probable error either. That is, there is a 50% chance that the actual error is less than the value we call the probable error. It is not the most probable error, but it is a measure of how large the error probably is.

The probable error has another significance. If we repeat the experiment, making the measurements in as nearly identical a manner as possible but not necessarily obtaining the identical observations, we expect the new result to have the same probable error as the first. Since we expect both determinations to be approximately within the probable error of the "true" value, they will also probably be within some fraction of the probable error of each other. Thus, the probable error for the result is also a measure of the probable discrepancy between two results obtained under identical conditions.

Notation   Since, in general, we will not be able to quote the actual error of the results, we must develop a consistent notation for quoting the estimated error. We must also realize that the model from which we calculate theoretical parameters to describe the results of our experiment may not be the "correct" model to use. In the chapters to come we will discuss hypothetical parameters and probable distributions of errors pertaining to the "true" state of affairs, and we will also discuss methods of making experimental estimates of these parameters and the probable errors corresponding to these determinations.

Minimizing uncertainties   Our preoccupation with error analysis is not confined just to a determination of the precision of our results. It is reasonable to expect that the most reliable results we can calculate from a given set of data will be those for which the estimated errors are the smallest. Thus, our development of techniques of error analysis will help to determine the optimum estimates of parameters to describe the data.

It must be noted, however, that our best results still will be only *estimates* of the quantities investigated.

SUMMARY

Systematic error: Reproducible inaccuracy introduced by faulty equipment, calibration, or technique.
Random error: Indefiniteness of result due to finite precision of experiment. Measure of fluctuation in result after repeated experimentation.

Probable error: Magnitude of error which is estimated to have been made in determination of results.

Accuracy: Measure of how close the result of the experiment comes to the "true" value.

Precision: Measure of how exactly the result is determined without reference to any "true" value.

Significant figures:

1. The leftmost nonzero digit is the most significant digit.
2. If there is no decimal point, the rightmost nonzero digit is the least significant digit.
3. If there is a decimal point, the rightmost digit is the least significant digit, even if it is a 0.
4. All digits between the least and most significant digits are counted as significant digits.

Round-off: Truncate the number to the specified number of significant digits and treat the excess digits as a decimal fraction.

1. If the fraction is greater than ½, increment the least significant digit.
2. If the fraction is less than ½, do not increment.
3. If the fraction equals ½, increment the least significant digit only if it is odd.

EXERCISES

1-1 How many significant figures are there in the following numbers?

    (*a*) 976.45   (*b*) 84,000   (*c*) 0.0094   (*d*) 301.07

    (*e*) 4.000   (*f*) 10   (*g*) 5280   (*h*) 400.

    (*i*) 4.00×10^2 (*j*) 3.010×10^4

1-2 What is the most significant digit in each of the numbers in Exercise 1-1? What is the least significant?

1-3 Round off each of the following numbers to two significant figures.

    (*a*) 4.451   (*b*) 1.75   (*c*) 2.045   (*d*) 1.03   (*e*) 1.85

    (*f*) 9.999   (*g*) 7.449   (*h*) 99.5   (*i*) 13.7   (*j*) 4.6675

1-4 Round off each of the numbers in Exercise 1-1 to two significant digits.

CHAPTER TWO

MEAN AND
STANDARD DEVIATION

2-1 PARENT POPULATION

If we make a measurement x_1 of a quantity x, we expect our observation to approximate the quantity, but we do not expect the experimental data point to be exactly equal to the quantity. If we make another measurement, we expect to observe a discrepancy between the two measurements due to random errors, and we don't expect either determination to be exactly correct, that is, equal to x. As we make more and more measurements, a pattern will emerge from the data. Some of the observations will be too large, some will be too small. On the average, however, we expect them to be distributed around the correct value, assuming we can neglect or correct for systematic errors.

If we were to make an infinite number of measurements, then we could describe the way in which the observed data points were distributed. This is not possible in practice, but we can hypothesize the existence of a similar probability distribution which determines the probability of getting any particular observation in one measurement. This distribution is called the *parent distribution*. Similarly, we can hypothesize that the measurements we have made are samples of an infinite number of possible measurements which are distributed according to the parent distribution. This collection of an infinite number of measurements is called the *parent population*.

In any practical experiment, we make a finite number of measurements. Our data represent a sample from the parent population, and the parent distribution indicates what that sample is probably like. Even if we cannot determine the quantity x exactly, we would like to be able to describe the parent distribution as well as possible.

Notation A number of parameters of the parent distribution have been defined by convention. We will use Greek letters to denote them, and Roman letters to denote experimental estimates of them.

In order to specify the parameters of the parent distribution, we will make use of the assumption that the results of experiments asymptotically approach the parent quantities as the number of measurements becomes infinitely large. We will denote this by saying that the parent parameters equal the experimental parameters *in the limit of an infinite number of measurements*. If we specify that there are N observations in a given experiment, then we can denote this with the notation

$$\text{(Parent parameter)} = \lim_{N \to \infty} \text{(experimental parameter)}$$

which states that in the limit that the number of observations N goes to infinity, the experimentally determined parameter becomes equal to the parameter of the parent distribution.

We will also make much use of the notation Σ to indicate the sum over a collection of items. If we make N measurements and label them x_1, x_2, etc., up to a final measurement x_N, then we can identify the sum of all these measurements as

$$\sum_{i=1}^{N} x_i \equiv x_1 + x_2 + x_3 + \cdots + x_N$$

where the left-hand side is interpreted as the sum of the observations x_i over the index i from $i = 1$ to $i = N$ inclusive. Since we will be making such frequent use of the sum over N measurements of various quantities, we will simplify the notation further and omit specific mention of the index whenever we are considering the sum where the index i runs from 1 to N.

$$\Sigma x_i \equiv \sum_{i=1}^{N} x_i$$

Mean and median   Using the definitions given above, the *mean* of the parent population μ is defined as the limit of the sum of N determinations x_i of the quantity x divided by the number N of determinations.

$$\mu \equiv \lim_{N \to \infty} \left(\frac{1}{N} \Sigma x_i \right) \tag{2-1}$$

The mean is therefore equivalent to the centroid or *average* value of the quantity x.

The *median* of the parent population $\mu_{1/2}$ is defined as that value for which, in the limit of an infinite number of determinations x_i of the quantity x, half the observations will be less than the median and half will be greater. In terms of the parent distribution, this means that the probability is 50% that any measurement x_i will be larger or smaller than the median

$$P(x_i \leq \mu_{1/2}) = P(x_i \geq \mu_{1/2}) = 50\%$$

so that the median line cuts the area of the probability distribution curve in half. Because of problems in computation, the median is no longer used very much as a statistical parameter.

The *most probable value* $\mu_{\max}$ of the parent population is that value for which the parent distribution has its greatest value. In

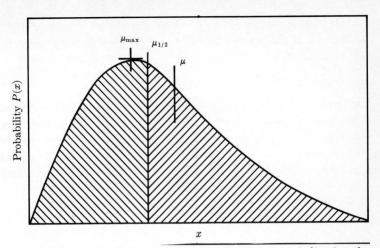

FIGURE 2-1 Typical probability distribution curve indicating the most probable value μ_{max}, the median $\mu_{1/2}$, and the mean μ. The two shaded areas are equal.

any given experimental measurement, this value is the one which is most likely to be observed. In the limit of a large number of observations, this value will probably occur most often.

$$P(\mu_{max}) \geq P(x \neq \mu_{max})$$

The relationship between the mean, median, and most probable value is illustrated on the probability distribution of Figure 2-1. For a symmetrical distribution, these parameters would all be equal by the symmetry of their definitions. For an asymmetric distribution such as that of Figure 2-1, the median generally falls between the most probable value and the mean. The most probable value corresponds to the peak of the distribution, and the areas on either side of the median are equal. For a true probability distribution, the magnitude is normalized so that the total area under the curve is equal to unity. The area under the curve at the point x bounded by a range dx is therefore equal to the probability $P(x) \, dx$ that a random sample differential measurement will yield an observed value of x.

Deviations The *deviation* d_i of any measurement x_i from the mean μ of the parent distribution is defined as the difference between x_i and μ.

$$d_i \equiv x_i - \mu$$

Deviations are generally defined with respect to the mean, rather than the median or most probable value for computational purposes. If μ is the true value of the quantity, d_i is also the true error in x_i.

The average of the deviations $\bar{d}$ for an infinite number of observations must vanish by virtue of the definition of the mean in Equation (2-1).

$$\lim_{N \to \infty} \bar{d} = \lim_{N \to \infty} \left[\frac{1}{N} \Sigma(x_i - \mu) \right] = \lim_{N \to \infty} \left(\frac{1}{N} \Sigma x_i \right) - \mu = 0$$

The *average deviation* α, therefore, is defined as the average of the magnitudes of the deviations, which are given by the absolute values of the deviations.

$$\alpha \equiv \lim_{N \to \infty} \left(\frac{1}{N} \Sigma |x_i - \mu| \right)$$

The average deviation is a measure of the *dispersion* of the expected observations about the mean. The presence of the absolute value signs makes it difficult to use for statistical analysis, however.

A parameter which is easier to use analytically and which can be justified fairly well on theoretical grounds (see Section 5-1) to be a more appropriate measure of the dispersion of observations is the standard deviation σ. The *variance* σ^2 is defined as the limit of the average of the squares of the deviations from the mean μ

$$\sigma^2 \equiv \lim_{N \to \infty} \left[\frac{1}{N} \Sigma(x_i - \mu)^2 \right] = \lim_{N \to \infty} \left(\frac{1}{N} \Sigma x_i^2 \right) - \mu^2 \qquad (2\text{-}2)$$

and the *standard deviation* σ is the square root of the variance. Thus, the standard deviation is the root mean square of the deviations, where we define the *root mean square* to be the square root of the mean or average of the square of an argument.

Significance   The mean μ, along with the median and most probable value, is a parameter which characterizes the information we are seeking when we perform an experiment. The mean may not be exactly equal to the datum in question if the parent distribution is not symmetrical about the mean, but it should have the same characteristics.

For example, in the experiment of Example 1-1, a student attempted to measure the length of a table. The information he sought was a parameter which described the length from one end to the other. From an atomic point of view, that length could not even theoretically be defined with infinite precision, and, in fact, the uncertainty in our result may reflect the physical uncertainty of the quantity being measured. But we can envision an ideal case where such an absolute "true" length might exist. If the parent distribution which determines the probability of observing particular lengths x_i is symmetrical about the mean, then the mean of the distribution is equal to the median and the most probable value, and these are all equal to the "true" length except for systematic errors. But if the distribution is not symmetrical about the mean, then we must know more about what causes the deviations to know whether the "true" length is characterized by the mean, the median, or the most probable value, if by any of these.

In general, the best we can say about the mean is that it is one of the parameters which specifies the probability distribution; it has the same units as the "true" value; and by convention, unless justified otherwise, we will consider it to be the best estimate that can be made of the "true" value under the prevailing experimental conditions.

The variance σ^2 and standard deviation σ characterize the uncertainties associated with our experimental attempts to determine the "true" values. There are two levels of abstraction associated with this uncertainty. Of primary importance is the relationship between the variance and our estimate from experimental data of the mean μ. For a given number of observations, the uncertainty in determining the mean of the parent distribution is proportional to the standard deviation of that distribution.

The standard deviation σ is therefore an appropriate measure of the uncertainty due to fluctuations in the observations.

At the same time we must remember that although the mean and "true" value may not be equal, their difference should be less than the uncertainty given by the probability distribution, and so the standard deviation should be a measure of the discrepancy between the parameter μ defined by a theoretical model and the "true" value.

Discrete distributions   We can define the mean μ and the standard deviation σ in terms of the parent distribution $P(x)$ of the parent population. The probability function $P(x)$ is defined such that in the limit that the number of observations N goes to infinity, the fraction of observations which yield a value of x is given by $P(x)$, which, in turn, must be the probability of obtaining a value of x in any single random observation.

For the definition of the mean μ in Equation (2-1), we can replace the sum over the individual observations Σx_i with a sum over the values of the possible observations times the number of times these observations are expected to occur. If there are n different possible observable values of the quantity x and we denote these by x_j (where the index j runs from 1 to n and no two values of x_j are equal), this sum can be expressed as

$$\lim_{N \to \infty} \Sigma x_i = \lim_{N \to \infty} \sum_{j=1}^{n} [x_j N P(x_j)]$$

The definition of the mean μ for a discrete distribution where the possible values of x are not continuous becomes

$$\mu = \lim_{N \to \infty} \sum_{j=1}^{n} [x_j P(x_j)] \qquad (2\text{-}3)$$

Similarly, the definition of the variance σ^2 in Equation (2-2) can be redefined in terms of the probability function $P(x)$.

$$\sigma^2 = \sum_{j=1}^{n} [(x_j - \mu)^2 P(x_j)] = \sum_{j=1}^{n} [x_j^2 P(x_j)] - \mu^2 \qquad (2\text{-}4)$$

The mean μ is the expectation value $\langle x \rangle$ of x, and the variance σ^2 is the expectation value $\langle (x - \mu)^2 \rangle$ of the square of the deviations of x from μ. The expectation value of any function $f(x)$ of x is defined as the weighted average value of the function averaged over all possible values of the variable x, with each value of $f(x)$ weighted by the probability distribution. It is given by

$$\langle f(x) \rangle = \sum_{j=1}^{n} [f(x_j)P(x_j)]$$

Continuous distributions If we consider the parent distribution as a continuous smoothly varying function $P(x)$ of the observed value x, assuming x can vary smoothly around μ, we can redefine μ and σ in terms of this probability function $P(x)$. The mean μ, as given in Equation (2-3), becomes the first moment of the parent distribution

$$\mu = \int_{-\infty}^{\infty} xP(x) \, dx \tag{2-5}$$

and the variance σ^2, as defined in Equation (2-4), becomes the second central product moment.

$$\sigma^2 = \int_{-\infty}^{\infty} (x - \mu)^2 P(x) \, dx = \int_{-\infty}^{\infty} x^2 P(x) \, dx - \mu^2 \tag{2-6}$$

2-2 SAMPLE PARAMETERS

In any practical experiment we will make only a finite number of observations. Let us consider that we are making N observations x_i (with i ranging from 1 to N) of a quantity which has a "true" value x_0. For our experimental conditions, we assume there exists a parent population of potential observations and a parent distribution which determines the probability of making any particular observation x_i. Our experimental group of N observations represents a sample of the parent population, and it is from this sample that all calculations must be made.

According to the development of Section 2-1, the experimental data will only suffice to enable us to *estimate* the param-

eters of the parent distribution of the parent population. We
will have to assume that these parameters give a valid description
of the "true" value x_0.

Sample mean For a series of N observations, the most
probable estimate of the mean μ is the average $\bar{x}$ of the observations (see Section 5-1). We will call this average the *sample mean*
$\bar{x}$ to distinguish it from the *parent mean* μ.

$$\mu \simeq \bar{x} \equiv \frac{1}{N} \Sigma x_i \tag{2-7}$$

The sum of the deviations $d_i = x_i - \bar{x}$ from the average $\bar{x}$ is 0
with this definition of $\bar{x}$. The average deviation (average of the
magnitudes) is not necessarily minimized, but it is easy to show
that the sum of the squares of the deviations d_i is minimized by
calculating the deviations with respect to the average $\bar{x}$. This
concept is the basis of the method of least squares to be discussed in later chapters.

To show that this is so, we expand the sum of squares of the
deviations

$$\Sigma(x_i - \bar{x})^2 = \Sigma x_i^2 - 2\bar{x}\Sigma x_i + N\bar{x}^2$$

and vary $\bar{x}$ until the expression is at a minimum. According to
the development of Appendix A, that value of $\bar{x}$ is one which
makes the derivative of the expression equal to 0. We take the
derivative with respect to the variable $\bar{x}$, noting that the experimental data points x_i are constants of the experiment and as such
have 0 derivatives (see Appendix A).

$$\frac{d}{dx} \Sigma(x_i - \bar{x})^2 = -2\Sigma x_i + 2N\bar{x} = 0 \tag{2-8}$$

Thus, the condition required to minimize the sum of squares of
deviations is the same as the definition we have already given to $\bar{x}$
as the average of the data points x_i; that is, Equation (2-8) is
equivalent to Equation (2-7).

Sample standard deviation We obtained our best estimate $\bar{x}$ of the mean μ of the parent distribution by using the

formula of Equation (2-1) for μ and taking only a finite sum instead of the limit of an infinite sum of terms. Similarly, our best estimate of the standard deviation σ would be obtained by modifying Equation (2-2).

$$\sigma^2 \simeq \frac{1}{N} \Sigma(x_i - \mu)^2 = \frac{1}{N} \Sigma x_i{}^2 - \mu^2 \qquad (2\text{-}9)$$

This formula presupposes an exact determination of the mean μ of the parent population, and we have at our disposal only an estimate of this parameter $\bar{x}$, the mean of the sample population. Since we arrived at this particular estimate of μ by minimizing the sum of squares of deviations, we might guess that the estimate formed by substituting $\bar{x}$ for μ in Equation (2-9) would be smaller than the estimate of Equation (2-9) before substitution.

We can compensate for this tendency to underestimate σ by modifying the form of Equation (2-9). The factor of N in the denominator represents the number of independent observations used to determine σ. It is the *number of degrees of freedom ν*, which is defined as the number of observations in excess of those needed to determine parameters appearing inside the equation. If we were to substitute $\bar{x}$ for μ in Equation (2-9), we would have to make at least one observation to determine $\bar{x}$. Instead of omitting that one observation from the sum in Equation (2-9), we actually use the data of all our observations to determine $\bar{x}$, so we decrease the number of degrees of freedom by 1. The best experimental estimate of the *parent standard deviation σ*, therefore, is given by the experimental or *sample standard deviation s*

$$\sigma^2 \simeq s^2 \equiv \frac{1}{N-1} \Sigma(x_i - \bar{x})^2 \qquad (2\text{-}10)$$

where $\nu = N - 1$ in the denominator is the number of degrees of freedom left after determining $\bar{x}$ from N observations.

We can justify this intuitively by considering our dilemma if we had made only one ($N = 1$) observation x_1. If we know the value of μ, we could use Equation (2-9) to estimate σ. But if we try to use $\bar{x}$ to estimate σ, we will find that the sum of squares of the deviations is equal to 0 by definition of $\bar{x} = x_1$. This would lead to an unreasonable estimate of σ except for the fact that the denominator of Equation (2-10) is also 0, and the experimental estimate is therefore indeterminate. As the number N of observations increases, our estimate of $\mu \simeq \bar{x}$ becomes more accurate, and the estimates for σ of Equations (2-9) and (2-10) become essentially equivalent.

Probability distributions What is the connection between the probability distribution of the parent population and the experimental sample we have taken? We have already seen that the uncertainties of the experimental conditions preclude a determination of the "true" values themselves. As a matter of fact, there are three levels of abstraction between the data and the information we seek:

1. From our experimental data points we can determine a sample frequency distribution which describes the way in which those particular data points are distributed over the range of possible data points. We use $\bar{x}$ to denote the mean of the data and $s^2(N - 1)/N$ to denote the variance. The shape and magnitude of the sample distribution are not exactly reproducible from sample to sample.
2. From the parameters of the sample probability distribution we can estimate the parameters of the parent probability distribution of the parent population of possible observations. Our best estimate for the mean μ is the mean of the sample distribution $\bar{x}$, and the best estimate for the variance σ^2 is the compensated variance s^2. Even the shape of this parent distribution must be estimated or assumed.
3. From the parameters of the parent distribution we estimate the results sought. In general, we will assume that some of the parameters of the parent distribution are equivalent to the "true" values, but the parent distribution is a function of the "true" values and the experimental conditions, and these may not necessarily be separable.

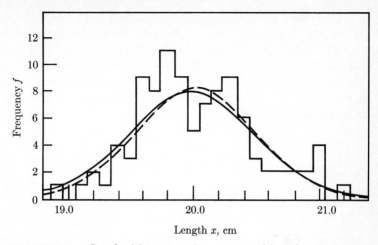

FIGURE 2-2 Graph of frequency f vs. measured length x for data of Table 2-1. Solid curve is actual parent distribution with $\mu = 20.000$ cm and $\sigma = 0.50$ cm. Dashed curve is estimated parent distribution with $\mu = 20.028$ cm and $\sigma = 0.48$ cm.

EXAMPLE 2-1 A student makes 100 measurements of the length of a wooden block. His observations, corrected for systematic errors, range from 18.9 cm to 21.2 cm, and many of the observations are identical. These measurements are listed in Table 2-1 as a frequency distribution indicating the number of times each length was observed. The mean of these data points is $\bar{x} = 20.03$ cm, and the standard deviation estimate is $s = 0.48$.

These data are graphed in Figure 2-2 as a *histogram* showing the number of observations recorded for each length. A histogram is a convenient method of graphing data to indicate simultaneously the smoothly varying nature of the parent distribution, of which the data represent a sample, and the fluctuating character of the finite collection of data points. The vertical portions of the outline indicate the separation of equal intervals of length; the horizontal portions indicate the number of observations falling within the corresponding interval. For comparison with theoretical models, the vertical bars should straddle the corresponding intervals. For example, in Figure 2-2 the observations were made

Table 2-1 Experimental data for the length of a block. These data are fabricated from a fictitious parent population with $\mu = 20.000$ cm and $\sigma = 0.500$ cm. The data tabulated are the number of times f each measurement x of the length of the block is observed

Length x, cm	Frequency f	fx	$x - \bar{x}$	$(x - \bar{x})^2$	$f(x - \bar{x})^2$
18.9	1	18.9	−1.128	1.2616	1.262
19.0	0	0.0	−1.028	1.0568	0.0
19.1	1	19.1	−0.928	0.8612	0.861
19.2	2	38.4	−0.828	0.6856	1.371
19.3	1	19.3	−0.728	0.5300	0.530
19.4	4	77.6	−0.628	0.3944	1.578
19.5	3	58.5	−0.528	0.2788	0.836
19.6	9	176.4	−0.428	0.1832	1.649
19.7	8	157.6	−0.328	0.1076	0.861
19.8	11	217.8	−0.228	0.0520	0.572
19.9	9	179.1	−0.128	0.0164	0.147
20.0	5	100.0	−0.028	0.0008	0.004
20.1	7	140.7	0.072	0.0052	0.036
20.2	8	161.6	0.172	0.0296	0.237
20.3	9	182.7	0.272	0.0740	0.666
20.4	6	122.4	0.372	0.1384	0.830
20.5	3	61.5	0.472	0.2228	0.668
20.6	2	41.2	0.572	0.3272	0.754
20.7	2	41.4	0.672	0.4516	0.903
20.8	2	41.6	0.772	0.5960	1.192
20.9	2	41.8	0.872	0.7604	1.521
21.0	4	84.0	0.972	0.9448	3.775
21.1	0	0.0	1.072	1.1492	0.0
21.2	1	21.2	1.172	1.3736	1.374
SUM	100	2002.8			22.627

$$\bar{x} = \frac{1}{N} \sum_{j=1}^{n} f_j x_j = \frac{2002.8}{100} = 20.028 \text{ cm}$$

$$s^2 = \frac{1}{N-1} \sum_{j=1}^{n} f_j (x_j - \bar{x})^2 = \frac{22.627}{99} = 0.229 \text{ cm}^2 \qquad s = \sqrt{s^2} = 0.48 \text{ cm}$$

to the nearest millimeter. The vertical lines separating the bars of the histograms, therefore, are drawn at the half-millimeter divisions.

Let us assume that over the years a large enough body of data has been collected to enable us to determine the parent distribution as well. Assuming a Gaussian distribution (see Section 3-3), this probability distribution is indicated as a solid line curve in Figure 2-2 for a mean $\mu = 20.00$ cm and a standard deviation $\sigma = 0.50$ cm. The data listed in Table 2-1 were actually derived from exactly this distribution, using a random number table and the Gaussian probability distribution to fabricate data points.

For comparison, the probability distribution corresponding to the estimates $\bar{x}$ and s of the data is indicated in Figure 2-2 as a dashed curve. Both probability distributions are normalized to correspond to a total of 100 data points for comparison with the experimentally determined histogram.

The difference between the experimental mean $\bar{x}$ and the "true" mean μ is quite apparent in Figure 2-2. Furthermore, it is not obvious that μ represents the true length x_0. Indeed, since the edge of the block is not perfectly smooth, we would have to define what we mean by the length x_0 anyway. We are restricted, in working with experimental data, to estimating the parameters of the parent population.

At the same time, by considering the data not as a collection of measurements of x_0 but as a sample of the parent population with the values of the observations distributed according to the parent population, we can make estimates of the shape and dispersion of the parent population, which give us useful information on the precision and reliability of our results. We estimate the mean μ in order to estimate the "true" value x_0; we also estimate the standard deviation σ in order to estimate the uncertainty of our result $\bar{x}$.

SUMMARY

Parent population: Hypothetical infinite set of data points of which the experimental data points are assumed to be a random sample.

Parent distribution: Probability distribution determining choice of sample data from the parent population.

Expectation value ⟨ ⟩: Weighted average of function $f(x)$ over all values of x.

$$\langle f(x) \rangle = \lim_{N \to \infty} \left[\frac{1}{N} \Sigma f(x_i) \right] = \sum_{j=1}^{n} [f(x_i)P(x_j)] = \int_{-\infty}^{\infty} f(x)P(x)\,dx$$

Median $\mu_{1/2}$:

$$P(x_i \leq \mu_{1/2}) = P(x_i \geq \mu_{1/2}) = \tfrac{1}{2}$$

Most probable value μ_{max}:

$$P(\mu_{max}) \geq P(x \neq \mu_{max})$$

Mean:

$$\mu \equiv \langle x \rangle$$

Average deviation:

$$\alpha \equiv \langle |x_i - \mu| \rangle$$

Variance:

$$\sigma^2 \equiv \langle (x_i - \mu)^2 \rangle = \langle x^2 \rangle - \mu^2$$

Standard deviation:

$$\sigma \equiv \sqrt{\sigma^2}$$

Sample mean:

$$\bar{x} = \frac{1}{N} \Sigma x_i$$

Sample variance:

$$s^2 = \frac{1}{N-1} \Sigma (x_i - \bar{x})^2$$

EXERCISES

2-1 Find the mean, median, and most probable value of x from the following data (from rolling dice).

i	x_i	i	x_i	i	x_i	i	x_i
1	8	6	6	11	8	16	11
2	10	7	5	12	5	17	12
3	9	8	6	13	8	18	6
4	5	9	3	14	10	19	7
5	9	10	9	15	8	20	8

2-2 Find the mean, median, and most probable grade from the following set of grades. Group them to find the most probable value.

i	x_i	i	x_i	i	x_i	i	x_i
1	49	11	90	21	69	31	74
2	80	12	84	22	69	32	86
3	84	13	59	23	53	33	78
4	73	14	56	24	55	34	55
5	89	15	62	25	77	35	66
6	78	16	53	26	82	36	60
7	78	17	83	27	81	37	68
8	92	18	81	28	76	38	92
9	56	19	65	29	79	39	87
10	85	20	81	30	83	40	86

2-3 Calculate the standard deviation for the data of Exercise 2-1.

2-4 Calculate the standard deviation for the data of Exercise 2-2.

2-5 The probability distribution for the sum of points showing on a pair of dice is given by

$$P(x) = \begin{cases} \dfrac{x-1}{36} & 2 \leq x \leq 7 \\ \dfrac{13-x}{36} & 7 \leq x \leq 12 \end{cases}$$

Find the mean, median, and standard deviation of the distribution.

2-6 The probability that an electron is at a distance r from the center of the nucleus of a hydrogen atom is given by

$$P(r) = Cr^2e^{-2r/R}$$

Find the mean radius $\bar{r}$ and the standard deviation.

2-7 Find the value of the constant C in Exercise 2-6.

2-8 Make histograms of the data of Exercises 2-1 and 2-2 and denote the means, medians, and standard deviations.

2-9 Justify the equality of Equations (2-2) and (2-9) that

$$\frac{1}{N} \Sigma(x_i - \mu)^2 = \frac{1}{N} \Sigma x_i^2 - \mu^2$$

2-10 Derive a relationship for the sum in Equation (2-10) similar to that of Exercise 2-9.

2-11 Justify the equality of Equations (2-4) and (2-6).

CHAPTER THREE

DISTRIBUTIONS

3-1 BINOMIAL DISTRIBUTION

If we toss a coin in the air and let it land, there is a 50% probability that it will land heads up and a 50% probability that it will land tails up. By this we mean that if we continue tossing the coin repeatedly, the fraction of times that it lands heads up will asymptotically approach $\frac{1}{2}$, indicating that there was a probability of $\frac{1}{2}$ of doing so. For any given toss, the probability cannot determine whether or not it will land heads up; it can only describe how we would expect a large number of tosses to be divided into the two possibilities.

Suppose we toss two coins at a time. There are now four different possible permutations of the way in which they can land: both heads up, both tails up, and two mixtures of heads and tails, depending on which one is heads up. Since each of these permutations is equally probable, the probability for any choice of them is $\frac{1}{4}$ or 25%. To find what the probability is for finding a mixture, without differentiating between the two kinds of mixtures, we must combine the probabilities corresponding to each possible kind by adding the probabilities. Thus, the probability for finding a mixture of one heads up and the other tails up is $\frac{1}{2}$. Note that the sum of the probabilities for all the possibilities $(\frac{1}{4} + \frac{1}{4} + \frac{1}{4} + \frac{1}{4})$ is always equal to 1 because *something* is bound to happen.

Let us extrapolate these ideas to the general case. Suppose we toss into the air n coins, where n is some integer. What is the probability that exactly x of these coins will land heads up, without distinguishing which of the coins actually belongs to which group? We can consider the probability $P(x,n)$ to be a function of the number n of coins tossed simultaneously and of the number x of coins which land heads up. For a given experiment in which a number n are tossed, this probability $P(x,n)$ will vary as a function of x. Of course, x must be an integer for any physical experiment, but we can consider the probability to be smoothly varying with x a continuous variable for mathematical purposes.

Permutations and combinations   If n coins are tossed, there are 2^n different possible ways in which they can land. This follows from the fact that the first coin has two possible orientations; for each of these the second coin also has two such orientations; for each of these the third coin has two; and so on. Since each of these possibilities is equally probable, the probability for any one of these possibilities to occur at any toss of n coins is $\frac{1}{2}^n$ or $(\frac{1}{2})^n$.

How many of these possibilities will contribute to our observations of x coins with heads up? Imagine two boxes, one labelled "heads" and divided into x slots, and the other labelled "tails." We will consider first the question of how many permutations of the coins result in the proper separation of x in one box and $n - x$ in the other, and then we will consider the question of how many combinations of these permutations should be considered different from each other.

In order to enumerate the number of *permutations* $Pm(n,x)$, let us pick up the coins one at a time from the collection of n coins and put x of them into the "heads" box. We have a choice of n coins for the first one we pick up. For our second selection we can choose from the remaining $n - 1$ coins. The range of choice is diminished until the last selection of the xth coin can be made from only $n - x + 1$ remaining coins. The total number of choices for coins to fill the x slots in the "heads" box is the product of the numbers of individual choices.

$$Pm(n,x) = n(n - 1)(n - 2) \cdots (n - x + 2)(n - x + 1)$$

This expansion can be expressed more easily in terms of factorials

$$P\dot{m}(n,x) = \frac{n!}{(n - x)!}$$

where the factorial function is defined (for integral values of n) as

$$n! \equiv n(n - 1)(n - 2) \cdots (3)(2)(1)$$
$$1! = 1 \qquad 0! = 1 \tag{3-1}$$

So far we have calculated the number of permutations $Pm(n,x)$ which will yield x coins in the "heads" box and $n - x$ coins in the "tails" box, with the provision that we have identified which coin was placed in the "heads" box first, which was placed second, and so on. That is, we have *ordered* the x coins in the "heads" box. In our computation of 2^n different possible permutations for the n coins, we only considered which coins landed heads up or down, not which one landed first. Therefore, we must consider contributions different only if there are different coins in the two boxes, not if the x coins within the "heads" box are permuted into different time sequences of choosing.

The number of different *combinations* $C(n,x)$ of the permutations enumerated above results from combining the $x!$ different ways in which x coins in the "heads" box can be permuted within the box. For every $x!$ permutations, there will be only one new combination. Thus, the number of different combinations $C(n,x)$

is the number of permutations $Pm(n,x)$ divided by the degeneracy factor $x!$ of the permutations.

$$C(n,x) = \frac{Pm(n,x)}{x!} = \frac{n!}{x!(n-x)!} \equiv \binom{n}{x}$$

This is the number of different combinations possible of n items taken x at a time, commonly referred to as $\binom{n}{x}$ or "n over x."

Probability The probability $P(x,n)$ that we would observe x coins with heads up and $n - x$ with tails up is the product of the number of different combinations $C(n,x)$ which contribute to that set of observations times the probability for each of the combinations to occur, which we have found to be $(\frac{1}{2})^n$.

Actually, we should separate the probability for each combination into two parts: one part is the probability $(\frac{1}{2})^x$ for x coins to be heads up; the other part is the probability $(\frac{1}{2})^{n-x}$ for the other $n - x$ coins to be tails up. The product of these two parts is the probability of the combination. In the general case (e.g., lopsided but identical coins), the probability p of success for each item (in this case landing heads up) is not equal to the probability $q = 1 - p$ for failure (landing tails up). The probability for each of the combinations of x coins heads up and $n - x$ coins tails up is $p^x q^{n-x}$.

With these definitions for p and q, the probability $P_B(x,n,p)$ for observing x of the n items to be in the state with probability p is given by the *binomial distribution*

$$P_B(x,n,p) = \binom{n}{x} p^x q^{n-x} = \frac{n!}{x!(n-x)!} p^x (1-p)^{n-x} \qquad (3\text{-}2)$$

where $q = 1 - p$. The name for the binomial distribution comes from the fact that the coefficients $P_B(x,n,p)$ are closely related to the binomial theorem for the expansion of a power of a sum. According to the binomial theorem,

$$(p+q)^n = \sum_{x=0}^{n} \left[\binom{n}{x} p^x q^{n-x} \right] \qquad (3\text{-}3)$$

Program 3-1 PBINOM Binomial probability distribution $P_B(x,n,p)$.

```
C      FUNCTION PBINOM
C
C      PURPOSE
C        EVALUATE BINOMIAL PROBABILITY COEFFICIENT
C
C      USAGE
C        RESULT = PBINOM (NOBS, NTOTAL, PROB)
C
C      DESCRIPTION OF PARAMETERS
C        NOBS    - NUMBER OF ITEMS OBSERVED
C        NTOTAL  - TOTAL NUMBER OF ITEMS
C        PROB    - PROBABILITY OF OBSERVING EACH ITEM
C
C      SUBROUTINES AND FUNCTION SUBPROGRAMS REQUIRED
C        FACTOR (N)
C            CALCULATES N FACTORIAL FOR INTEGERS
C
C      MODIFICATIONS FOR FORTRAN II
C        NONE
C
       FUNCTION PBINOM (NOBS, NTOTAL, PROB)
     1 NOTOBS = NTOTAL - NOBS
     2 PBINOM = FACTOR(NTOTAL) / (FACTOR(NOBS)*FACTOR(NOTOBS))
     1 * (PROB**NOBS) * (1.-PROB)**NOTOBS
       RETURN
       END
```

The $(j + 1)$th term, corresponding to $x = j$, of this expansion, therefore, is equal to the probability $P_B(j,n,p)$. We can use this result to show that the binomial distribution coefficients $P_B(x,n,p)$ are normalized to a sum of 1, that is, the sum of probabilities for all values of x from 0 to n is unity. The right-hand side of Equation (3-3) is the sum of the probabilities over all possible values of x from 0 to n and the left-hand side is $1^n = 1$.

Program 3-1 The calculation of Equation (3-2) is illustrated in the computer routine PBINOM of Program 3-1. This is a Fortran function subprogram to calculate the binomial coefficient $P_B(x,n,p)$ for a given value of x, n, and p. The input variables are NOBS = x (number observed), NTOTAL = n (number of possible values for x), and PROB = p (probability of success). The variable NOTOBS = $n - x$ is defined in statement 1. $P_B(x,n,p)$ is calculated in statement 2 according to Equation (3-2) and returned to the main program as the value of the function PBINOM.

Program 3-2   FACTOR   Factorial function $n!$

```
C       FUNCTION FACTOR
C
C       PURPOSE
C         CALCULATES FACTORIAL FUNCTION FOR INTEGERS
C
C       USAGE
C         RESULT = FACTOR (N)
C
C       DESCRIPTION OF PARAMETERS
C         N        - INTEGER ARGUMENT
C
C       SUBROUTINES AND FUNCTION SUBPROGRAMS REQUIRED
C         NONE
C
C       MODIFICATIONS FOR FORTRAN II
C         OMIT DOUBLE PRECISION SPECIFICATIONS
C         CHANGE DLOG TO LOGF IN STATEMENT 34
C         CHANGE DEXP TO EXPF IN STATEMENT 35
C
        FUNCTION FACTOR (N)
        DOUBLE PRECISION FI, SUM
     11 FACTOR = 1.
        IF (N-1) 40, 40, 13
     13 IF (N-10) 21, 21, 31
C
C           N LESS THAN 11
C
     21 DO 23 I=2, N
        FI = I
     23 FACTOR = FACTOR * FI
        GO TO 40
C
C           N GREATER THAN 10
C
     31 SUM = 0.
        DO 34 I=11, N
        FI = I
     34 SUM = SUM + DLOG(FI)
     35 FACTOR = 3628800. * DEXP(SUM)
     40 RETURN
        END
```

Program 3-2   The factorial function is evaluated with the computer routine FACTOR of Program 3-2. This is a Fortran function subprogram to calculate FACTOR(N) = $N!$ as defined in Equation (3-1). For N less than 11, the computation is a straightforward multiplication in statements 21–23. Logarithms are used in statements 31–35 for larger values of N to preserve precision.

Mean and standard deviation The mean μ of a binomial distribution is evaluated by combining the definition of μ in Equation (2-3) with the formula for the probability function of Equation (3-2) (see Exercise 3-12 for proof).

$$\mu = \sum_{x=0}^{n} \left[x \frac{n!}{x!(n-x)!} p^x (1-p)^{n-x} \right] = np \tag{3-4}$$

This means that if we perform an experiment with n items and observe the number x of successes, after a large number of repeated experiments the average $\bar{x}$ of the number of successes will approach a mean value μ given by the probability for success of each item p times the number of items n. In the case of coin tossing where $p = \frac{1}{2}$, we would expect on the average to observe half the coins land heads up, which seems eminently reasonable.

The variance σ^2 of a binomial distribution is similarly evaluated by combining Equations (2-4) and (3-2) (see Exercise 3-13 for proof).

$$\sigma^2 = \sum_{x=0}^{n} \left[(x-\mu)^2 \frac{n!}{x!(n-x)!} p^x (1-p)^{n-x} \right]$$
$$= np(1-p) \tag{3-5}$$

The evaluation of these sums is left as an exercise for the reader. We are mainly interested in the results which are remarkably simple.

If the probability for a single success p is equal to the probability for failure $p = q = \frac{1}{2}$, the distribution is symmetric about the mean μ, and the median $\mu_{\frac{1}{2}}$ and the most probable value are both equal to the mean. In this case the variance σ^2 is equal to half the mean $\sigma^2 = \mu/2$. If p and q are not equal, the distribution is asymmetric with a smaller variance.

EXAMPLE 3-1 Suppose we toss 10 coins in the air a total of 100 times. With each toss we observe the number of coins that

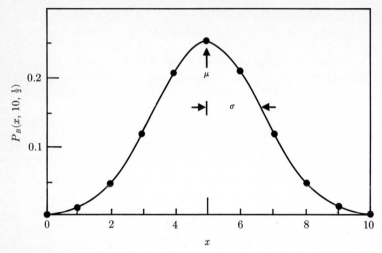

FIGURE 3-1 Binomial distribution function $P_B(x,n,p)$ for $n = 10$ and $p = \frac{1}{2}$; $\mu = 5$ and $\sigma = 1.6$.

land heads up and denote that number by x_i where i is the number of the toss; i ranges from 1 to 100 and x_i can be any integer from 0 to 10. The probability function governing the distribution of the observed values of x is given by the binomial distribution $P_B(x,n,p)$ with $n = 10$ and $p = \frac{1}{2}$. The parent distribution is not affected by the number N of repeated procedures in the experiment.

The parent distribution $P_B(x,10,\frac{1}{2})$ is shown in Figure 3-1 as a smooth curve drawn through discrete points. Only the points have physical significance, but the smooth curve helps to illustrate the functional behavior of the binomial distribution for large values of n. The mean μ is given by Equation (3-4).

$$\mu = 10(\tfrac{1}{2}) = 5$$

The standard deviation σ is given by Equation (3-5).

$$\sigma = \sqrt{10(\tfrac{1}{2})(\tfrac{1}{2})} = \sqrt{2.5} \simeq 1.58$$

The curve is symmetric about its peak at the mean and the magnitudes of the points are such that the sum of the probabilities over all of the points is equal to 1.

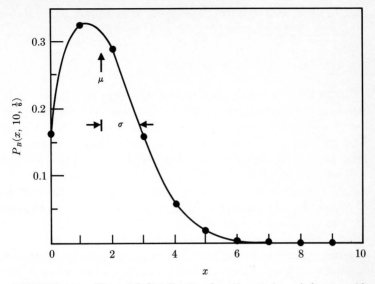

FIGURE 3-2 Binomial distribution function $P_B(x,n,p)$ for $n = 10$ and $p = \frac{1}{6}$; $\mu = 1.67$ and $\sigma = 1.2$.

EXAMPLE 3-2   Suppose we roll 10 dice. What is the probability that x of these dice will land with 1 up? If we throw one die, the probability of it landing with 1 up is $p = \frac{1}{6}$. If we throw 10 dice, the probability for x of them landing with 1 up is given by the binomial distribution $P_B(x,n,p)$ with $n = 10$ and $p = \frac{1}{6}$.

$$P_B\left(x,10,\frac{1}{6}\right) = \frac{10!}{x!(10-x)!}\left(\frac{1}{6}\right)^x \left(\frac{5}{6}\right)^{10-x}$$

This distribution is illustrated in Figure 3-2 as a smooth curve drawn through discrete points as in Figure 3-1. The mean and standard deviation are $\mu = \frac{10}{6} \simeq 1.67$ and

$$\sigma = \sqrt{10(\frac{1}{6})(\frac{5}{6})} = \frac{5}{6}\sqrt{2} \simeq 1.18$$

The distribution is not symmetric about the mean or about any other point. The most probable value is for $x = 1$, but the peak of the smooth curve occurs for a slightly larger value of x. Note that the standard deviation σ is smaller than that of Example 3-1 even though the number of items n is the same.

3-2 POISSON DISTRIBUTION

The Poisson distribution represents an approximation to the binomial distribution for the special case when the average number of successes is very much smaller than the possible number; that is, when $\mu \ll n$ because $p \ll 1$. Such an approximation is generally appropriate for counting experiments where the data represent the number of items observed per unit time interval.

For example, in the study of radioactive decay, the number of nuclei in a milligram of radioactive material is on the order of 10^{19}. In an experiment designed to determine the decay rate, the number of nuclei which disintegrate in each time interval is many orders of magnitude smaller. Similarly, in a study of nuclear reactions with an accelerator, the beam incident on a target may be comprised of on the order of 10^{13} particles per second while the number of events recorded in experimental detectors is less than 10^4 per second.

For such experiments the binomial distribution correctly describes the probability $P_B(x,n,p)$ of observing x events per unit time interval out of n possible events, each of which has a probability p of occurring, but the larger number n of possible events makes exact evaluation impossible. Furthermore, in the experiments described above, neither the number n of possible events nor the probability p for each is generally known. What may be known instead is the average number of events observed per unit time μ or its estimate $\bar{x}$. The Poisson distribution provides an analytical form appropriate to such investigations which describes the probability distribution in terms of just the parameters x and μ.

Binomial distribution for $p \ll 1$ Let us consider the binomial distribution in the limiting case of $p \ll 1$. We are interested in its behavior as n becomes infinitely large while the mean $\mu = np$ remains constant. Equation (3-2) for the probability function of the binomial distribution may be written as

$$P_B(x,n,p) = \frac{1}{x!} \frac{n!}{(n-x)!} p^x (1-p)^{-x} (1-p)^n$$

If we expand the second term

$$\frac{n!}{(n-x)!} = n(n-1)(n-2) \cdots (n-x+2)(n-x+1)$$

we can consider it to be the product of x terms, each of which is very nearly equal to n since $x \ll n$ in the region of interest. This term asymptotically approaches n^x. The product of the second and third terms becomes $(np)^x = \mu^x$. The fourth term is approximately equal to $1 + px$ which asymptotically approaches 1 as p becomes infinitesimally small.

The last term can be rearranged by substituting μ/p for n to show that it asymptotically approaches $e^{-\mu}$.

$$\lim_{p \to 0} (1-p)^n = \lim_{p \to 0} [(1-p)^{1/p}]^\mu = \left(\frac{1}{e}\right)^\mu = e^{-\mu}$$

Combining these approximations, we find that the binomial distribution probability function $P_B(x,n,p)$ asymptotically approaches the *Poisson distribution* $P_P(x,\mu)$.

$$\lim_{p \to 0} P_B(x,n,p) = P_P(x,\mu) \equiv \frac{\mu^x}{x!} e^{-\mu} \tag{3-6}$$

Since this distribution is an approximation to the binomial distribution for $p \ll 1$, the distribution will resemble that of Figure 3-2, asymmetric about its mean μ. Note that $P_P(x,\mu)$ does not go to 0 for $x = 0$, but it is undefined for negative values of x. This restriction is not troublesome for counting experiments because the number of counts per unit time interval can never be negative.

In order to verify that the normalization of this distribution is correct, we note that the sum over the first term of Equation (3-6) is just the series expansion of the exponential

$$\sum_{x=0}^{\infty} \frac{\mu^x}{x!} = e^\mu$$

where the sum is taken over an infinite number of terms because the Poisson distribution corresponds to the limit of the binomial distribution as n becomes infinitely large. The sum of the probability $P_P(x,\mu)$ over all x therefore equals 1.

Exact derivation   The Poisson distribution can also be derived exactly for the case where the sample of events observed is small compared to the total possible number of events.[1] Suppose, for example, that the average rate at which events of interest occur is constant over a given range of time and that the occurrences of the events are randomly distributed over that range. The probability of observing no events within a time t is then given by

$$P(0,t,\tau) = e^{-t/\tau}$$

where τ is a constant proportionality factor which may be associated with the mean time between events. This functional behavior follows from the assumption of randomicity that increasing the time interval t by a differential dt decreases the probability proportionally.

$$dP(0,t,\tau) = -P(0,t,\tau)\,\frac{dt}{\tau}$$

The above assumptions are analogous to those of the previous derivation because they assume no constraints on the number of events which may be observed, and this implies that the average number of events observed $\mu = np$ is much smaller than the maximum number of events n which could be observed.

The probability $P(x,t,\tau)$ for observing x events in the time interval t can be evaluated by integrating the differential probability $d^x P(x,t,\tau)$, which is the product of the probabilities of observing each event in a different differential interval dt_i times

[1] This derivation of the Poisson distribution follows that of Orear, pp. 21–22.

Program 3-3   PPOISS   Poisson probability distribution $P_P(x,\mu)$.

```
C       FUNCTION PPOISS
C
C       PURPOSE
C          EVALUATE POISSON PROBABILITY FUNCTION
C
C       USAGE
C          RESULT = PPOISS (NOBS, AVERAG)
C
C       DESCRIPTION OF PARAMETERS
C          NOBS   - NUMBER OF ITEMS OBSERVED
C          AVERAG - MEAN OF DISTRIBUTION
C
C       SUBROUTINES AND FUNCTION SUBPROGRAMS REQUIRED
C          FACTOR (N)
C             CALCULATES N FACTORIAL FOR INTEGERS
C
C       MODIFICATIONS FOR FORTRAN II
C          ADD F SUFFIX TO EXP IN STATEMENT 1
C
        FUNCTION PPOISS (NOBS, AVERAG)
      1 PPOISS = ((AVERAG**NOBS)/FACTOR(NOBS)) * EXP (-AVERAG)
        RETURN
        END
```

the probability $e^{-t/\tau}$ of not observing any other events in the remaining time.

$$d^x P(x,t,\tau) = e^{-t/\tau} \frac{1}{x!} \prod_{i=1}^{x} \frac{dt_i}{\tau}$$

The factor of $1/x!$ compensates for the ordering implicit in the product of probabilities $dP_i(1,t,\tau)$ as in the discussion of the combinations $C(n,x)$ in Section 3-1.

Thus, the probability of observing x events in the time interval t is the same as that of Equation (3-6)

$$P_P(x,\mu) = P(x,t,\tau) = e^{-t/\tau} \frac{1}{x!} \left(\frac{t}{\tau}\right)^x = \frac{\mu^x}{x!} e^{-\mu}$$

where $\mu = t/\tau$ is the average number of events observed in the time interval t.

Program 3-3   The distribution of Equation (3-6) is illustrated in the computer routine ppoiss of Program 3-3. This is a Fortran function subprogram to calculate the Poisson probability

$P_P(x,\mu)$ for a given value of x and μ. The input variables are NOBS $= x$ (number observed) and AVERAG $= \mu$ (mean of distribution). The probability $P_P(x,\mu)$ is calculated in statement 1 according to Equation (3-6) and returned to the main program as the value of the function PPOISS. Factorials are evaluated with the function FACTOR of Program 3-2.

Mean and standard deviation The mean μ of the Poisson distribution must be the same parameter μ which appears in the probability function $P_P(x,\mu)$ of Equation (3-6) because that is the mean of the corresponding binomial distribution. To verify this, we can evaluate the expectation value $\langle x \rangle$ of x.

$$\langle x \rangle = \sum_{x=0}^{\infty} \left(x \frac{\mu^x}{x!} e^{-\mu} \right) = \mu e^{-\mu} \sum_{x=1}^{\infty} \frac{\mu^{x-1}}{(x-1)!} = \mu e^{-\mu} \sum_{y=0}^{\infty} \frac{\mu^y}{y!} = \mu$$

The variance σ^2 can be evaluated by examining the behavior of the variance of the binomial distribution given in Equation (3-5) in the limit that p becomes infinitesimally small.

$$\sigma^2 = \lim_{p \to 0} [np(1-p)] = \mu \tag{3-7}$$

Alternatively, the expectation value of the square of deviations can be evaluated directly (see Exercise 3-14 for proof).

$$\sigma^2 = \langle (x - \mu^2) \rangle = \sum_{x=0}^{\infty} \left[(x - \mu)^2 \frac{\mu^x}{x!} e^{-\mu} \right] = \mu \tag{3-8}$$

The standard deviation σ, therefore, is equal to the square root of the mean μ.

EXAMPLE 3-3 A detector is placed near a weak radioactive source and the number of counts in the detector is recorded at 10-sec intervals. After considerable experimentation it is established that the distribution of the number of counts observed per interval is given by the Poisson distribution with a mean

$\mu = 1.67$ counts/10 sec

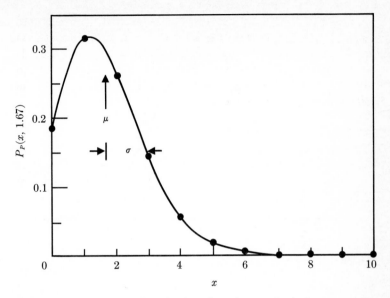

FIGURE 3-3 Poisson distribution function $P_P(x,\mu)$ for $\mu = 1.67$; $\sigma = 1.29$.

This distribution is shown in Figure 3-3 as a smooth curve drawn through discrete points, as in Figure 3-1.

The asymmetry of the distribution is quite noticeable, as is the fact that the mean μ does not coincide with the most probable value of x at the peak of the curve. For discrete values of x, the most probable value is $x = 1$, not $x = 2$ as we might have expected from the fact that the mean is closer to 2. The distribution of Figure 3-3 is very similar to that of Figure 3-2, indicating that p does not have to be very small for the approximation to be useful.

The standard deviation is given by Equations (3-7) and (3-8): $\sigma = \sqrt{\mu} = \sqrt{1.67} = 1.29$. This value is only about 10% higher than that calculated for Example 3-2. A comparison of σ and μ indicates that with such a low counting rate we would have difficulty in determining the average counting rate with very good precision.

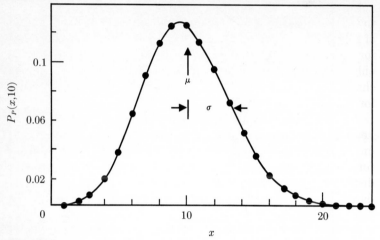

FIGURE 3-4 Poisson distribution function $P_P(x,\mu)$ for $\mu = 10$; $\sigma = 3.16$.

EXAMPLE 3-4   Suppose that in order to improve the experiment of Example 3-3 we make measurements over 1-min intervals. The average counting rate will be six times higher:

$$\mu = 10 \text{ counts/min}$$

The standard deviation is increased by a factor of $\sqrt{6}$:

$$\sigma = \sqrt{10} = 3.16$$

The Poisson distribution for these conditions is shown in Figure 3-4. The distribution is more nearly symmetric with the mean and the peak of the curve about ½ count/min apart. As the mean μ becomes considerably larger than 1, the Poisson distribution becomes more nearly symmetric.

Significance   A comparison of the curves of Figures 3-1 to 3-4 reveals the nature of the Poisson distribution. It is the appropriate distribution for describing experiments in which the possible values of the data are strictly bounded on one side but not on the other. The curve of Figure 3-1 is bounded equally close to the mean on both sides and is therefore symmetric. The curve of Figure 3-2 is bounded more closely on the left side and is therefore asymmetric. The curve of Figure 3-3 is bounded essentially only on the left side and exhibits true Poisson behavior. It is this characteristic asymmetry which is usually considered appropriate for the Poisson distribution.

The Poisson distribution is equally valid in describing the nearly symmetric curve of Figure 3-4 which is also bounded only on the left side. The advantage of using the Poisson distribution for analyzing any such experiment comes from the fact that the distribution is specified completely by the mean μ, and the standard deviation σ is uniquely determined by that mean.

3-3   GAUSSIAN OR NORMAL ERROR DISTRIBUTION

The most important probability distribution for use in statistical analysis of data is the Gaussian or normal error distribution. Mathematically, it is an approximation to the binomial distribution for the special limiting case where the number of possible different observations n becomes infinitely large and the probability of success for each is finitely large so that $np \gg 1$. Physically, however, it is useful because it seems to describe the distribution of random observations for most experiments where the Poisson distribution is not appropriate, and, furthermore, it appears to describe well the distribution of estimations of the parameters of most probability distributions.

There are a number of derivations of the Gaussian distribution from first principles, none of them as convincing as the fact that the distribution is reasonable, that it has a fairly simple analytic form, and that it is accepted by convention and experimentation to be the most likely distribution for most experiments. In addition, it does have the satisfying characteristic that the most probable estimate of the mean μ from a random sample of observations x is the average of those observations $\bar{x}$.

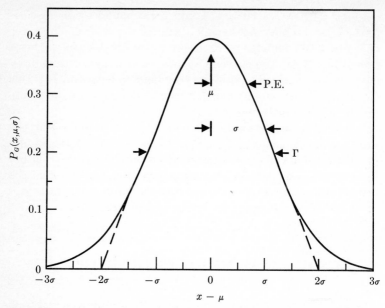

FIGURE 3-5 Gaussian probability distribution $P_G(x,\mu,\sigma)$ vs. $x - \mu$; $\Gamma = 2.345\sigma$; P.E. $= 0.6745\sigma$.

Characteristics The Gaussian distribution function is defined as

$$P_G(x,\mu,\sigma) = \frac{1}{\sigma \sqrt{2\pi}} \exp\left[-\frac{1}{2}\left(\frac{x - \mu}{\sigma}\right)^2 \right] \tag{3-9}$$

It is a continuous function describing the probability that from a parent distribution with a mean μ and a standard deviation σ, the value of a random observation would be x. Since the distribution is continuous, we must define an interval in which the value of the observation x will fall. The probability function is properly defined such that the probability $dP_G(x,\mu,\sigma)$ that the value of a random observation will fall within an interval dx around x is given by

$$dP_G(x,\mu,\sigma) = P_G(x,\mu,\sigma)\ dx$$

considering dx to be an infinitesimal differential.

Program 3-4 PGAUSS Gaussian probability distribution $P_G(x,\mu,\sigma)$.

```
C     FUNCTION PGAUSS
C
C     PURPOSE
C       EVALUATE GAUSSIAN PROBABILTY FUNCTION
C
C     USAGE
C       RESULT = PGAUSS (X, AVERAG, SIGMA)
C
C     DESCRIPTION OF PARAMETERS
C       X      - VALUE FOR WHICH PROBABILITY IS TO BE EVALUATED
C       AVERAG - MEAN OF DISTRIBUTION
C       SIGMA  - STANDARD DEVIATION OF DISTRIBUTION
C
C     SUBROUTINES AND FUNCTION SUBPROGRAMS REQUIRED
C       NONE
C
C     MODIFICATIONS FOR FORTRAN II
C       OMIT DOUBLE PRECISION SPECIFICATIONS
C       CHANGE DEXP TO EXPF IN STATEMENT 2
C
      FUNCTION PGAUSS (X, AVERAG, SIGMA)
      DOUBLE PRECISION Z
    1 Z = (X-AVERAG)/SIGMA
    2 PGAUSS = 0.3989422804/SIGMA * DEXP(-(Z**2)/2.)
      RETURN
      END
```

The shape of the distribution is illustrated in Figure 3-5. It is a bell-shaped curve symmetric about the mean μ. The width of the curve is characterized by the *full-width at half-maximum*, which is generally called the *half-width* Γ. This is defined as the range of x between the values at which the probability $P_G(x,\mu,\sigma)$ is half its maximum value.

$$P_G(\mu \pm \tfrac{1}{2}\Gamma, \mu, \sigma) = \tfrac{1}{2}P_G(\mu,\mu,\sigma) \qquad \Gamma = 2.354\sigma \qquad (3\text{-}10)$$

A tangent drawn along the portion of steepest descent of the curve intersects the curve at the $e^{-\frac{1}{2}}$ points $x = \mu \pm \sigma$ and intersects the x axis at the points $x = \mu \pm 2\sigma$.

$$P_G(\mu \pm \sigma, \mu, \sigma) = e^{-\frac{1}{2}}P_G(\mu,\mu,\sigma)$$

Program 3-4 The calculation of Equation (3-9) is illustrated in the computer routine PGAUSS of Program 3-4. This is a Fortran function subprogram to calculate the Gaussian distribution function $P_G(x,\mu,\sigma)$ for a given value of x, μ, and σ. The input variables are $X = x$ (observed value), AVERAG $= \mu$ (mean of distribution), and SIGMA $= \sigma$ (standard deviation).

The variable $Z = (x - \mu)/\sigma$ evaluated in statement 1 is the dimensionless ratio of the deviation of x from μ divided by the standard deviation σ. The probability function $P_G(x,\mu,\sigma)$ is calculated in statement 2 and returned to the main program as the value of the function PGAUSS.

Mean and standard deviation   The mean and standard deviation of the Gaussian distribution are equal to the parameters μ and σ given explicitly in the form of Equation (3-9) for the distribution function. This equivalence can be verified by using the definitions of μ and σ as the expectation values of x and $(x - \mu)^2$ for continuous distributions as given in Equations (2-5) and (2-6).

The *probable error* P.E. is defined to be the absolute value of the deviation $|x - \mu|$ such that the probability for the deviation of any random observation $|x_i - \mu|$ to be less is equal to $\frac{1}{2}$. That is, half the observations of an experiment are expected to fall within the boundaries denoted by $\mu \pm$ P.E. The relation between the P.E. and the standard deviation σ is found by evaluating the point at which the integral probability curve (see below) yields a probability of $\frac{1}{2}$ of being exceeded.

$$\text{P.E.} = .6745\sigma = .2865\Gamma \tag{3-11}$$

Integral probability   The integral probability $A_G(x,\mu,\sigma)$ is defined as the probability that the value of any random measurement x_i will have a deviation from μ less than $z\sigma$, where z is the dimensionless range $z = |x - \mu|/\sigma$. Therefore, it is equal to the integral (or area) of the probability function $P_G(x,\mu,\sigma)$ evaluated between the limits $\mu \pm z\sigma$.

$$A_G(x,\mu,\sigma) = \int_{\mu-z\sigma}^{\mu+z\sigma} P_G(x,\mu,\sigma)\, dx = \frac{1}{\sqrt{2\pi}} \int_{-z}^{z} e^{-\frac{1}{2}x^2}\, dx \tag{3-12}$$

The normalization of the probability function $P_G(x,\mu,\sigma)$ is such that the integral over all values of x is equal to unity:

$$A_G(z = \infty) = 1$$

Program 3-5 Values of the integral of Equation (3-12) cannot be evaluated easily because there is no analytical form for the integral. To evaluate it we must expand the exponential and integrate.

$$
\begin{aligned}
A_G(x,\mu,\sigma) &= \frac{2}{\sqrt{\pi}} \int_0^{z/\sqrt{2}} e^{-y^2}\, dy \\
&= \frac{2}{\sqrt{\pi}} \int_0^{z/\sqrt{2}} \left(1 - y + \frac{y^2}{2!} - \frac{y^3}{3!} + \cdots \right) dy \\
&= \frac{2}{\sqrt{\pi}} \sum_{k=0}^{\infty} \frac{(-1)^k (z/\sqrt{2})^{2k+1}}{k!(2k+1)} \\
&= \frac{2}{\sqrt{\pi}} e^{-\frac{1}{2}z^2} \sum_{k=0}^{\infty} \frac{(z/\sqrt{2})^{2k+1} 2^k}{(2k+1)!!}
\end{aligned}
\tag{3-13}
$$

where the double factorial means

$$
(2k+1)!! \equiv (2k+1)(2k-1)(2k-3) \cdots (3)(1)
$$

Of the two expressions for $A_G(x,\mu,\sigma)$ in Equation (3-13), the second is preferable for computation purposes because it converges faster and because the errors from subtracting nearly equal terms in the first expression have been reduced. This calculation is illustrated in the computer routine AGAUSS of Program 3-5. This is a Fortran function subprogram to calculate the integral of the Gaussian probability distribution $A_G(x,\mu,\sigma)$ for a given value of $z = |x - \mu|/\sigma$. The input variables $X = x$, AVERAG $= \mu$, and SIGMA $= \sigma$ are defined as in Program 3-4.

The dimensionless range $z = |x - \mu|/\sigma$ is evaluated in statement 11. The first term of the sum is computed in statement 21. Succeeding terms are computed in statements 31–32 and the sum is accumulated in statement 33. If the last term is large enough to affect the sum, the calculation and accumulation of the next term is continued at statement 31. Otherwise, the result is evaluated in statement 41 and returned to the main program as the value of the function AGAUSS.

Program 3-5 AGAUSS Integral Gaussian probability function $A_G(x,\mu,\sigma)$.

```
C       FUNCTION AGAUSS
C
C       PURPOSE
C         EVALUATE INTEGRAL OF GAUSSIAN PROBABILITY FUNCTION
C
C       USAGE
C         RESULT = AGAUSS (X, AVERAG, SIGMA)
C
C       DESCRIPTION OF PARAMETERS
C         X       - LIMIT FOR INTEGRAL
C         AVERAG - MEAN OF DISTRIBUTION
C         SIGMA   - STANDARD DEVIATION OF DISTRIBUTION
C         INTEGRATION RANGE IS AVERAG +/- Z*SIGMA
C             WHERE Z = ABS(X-AVERAG)/SIGMA
C
C       SUBROUTINES AND FUNCTION SUBPROGRAMS REQUIRED
C         NONE
C
C       MODIFICATIONS FOR FORTRAN II
C         OMIT DOUBLE PRECISION SPECIFICATIONS
C         ADD F SUFFIX TO ABS IN STATEMENT 11
C         CHANGE DEXP TO EXPF IN STATEMENT 41
C
        FUNCTION AGAUSS (X, AVERAG, SIGMA)
        DOUBLE PRECISION Z, Y2, TERM, SUM, DENOM
     11 Z = ABS (X-AVERAG)/SIGMA
        AGAUSS = 0.
        IF (Z) 42, 42, 21
     21 TERM = 0.7071067812 * Z
     22 SUM = TERM
        Y2 = (Z**2)/2.
        DENOM = 1.
C
C          ACCUMULATE SUM OF TERMS
C
     31 DENOM = DENOM + 2.
     32 TERM = TERM * (Y2*2./DENOM)
     33 SUM = SUM + TERM
        IF (TERM/SUM - 1.E-10) 41, 41, 31
     41 AGAUSS = 1.128379167 * SUM * DEXP(-Y2)
     42 RETURN
        END
```

Tables and graphs The Gaussian probability function $P_G(x,\mu,\sigma)$ and the integral probability $A_G(x,\mu,\sigma)$ are tabulated and graphed in Appendixes C-1 and C-2, respectively. The values tabulated were calculated with the functions of Programs 3-4 and 3-5. The functions are tabulated and graphed as functions of the dimensionless deviation $z = |x - \mu|/\sigma$.

EXAMPLE 3-5 Consider again the experiment of Example 2-1. The distribution of the observed measurements of the length of the block is similar to the Gaussian distribution as illustrated in Figure 2-1. Presumably, if enough data points were obtained, the experimental distribution could be made to simulate the parent Gaussian distribution to any desired precision.

The theoretical parameters for the parent distribution of this experiment were specified as $\mu = 20.000$ cm and $\sigma = 0.500$ cm. The experimentally determined sample parameters $\bar{x} = 20.03$ cm and $s = 0.48$ cm are good approximations to these values. Note that all but two of the data points fall within 2 standard deviations of the mean. This is essentially what we would expect from the fact that the probability of exceeding 2 standard deviations is less than 5% (see Table C-3). Out of 100 observations, we would only expect a few values to deviate by that much. Similarly, the probability of exceeding 3 standard deviations is less than 0.3%. In general, unless there are over 100 observations, we may be suspicious of any observation deviating this much from the mean.

The P.E. for the parent distribution is calculated from Equation (3-11): P.E. = $.6745\sigma$ = $.337$ cm. As expected, about half the observations fall between 19.7 and 20.3 cm. If we made one additional measurement of the length, there would be a 50% chance that its value would also fall within these limits.

3-4 LORENTZIAN DISTRIBUTION

There is another distribution which is similar but unrelated to the binomial distribution which occurs quite often in nuclear physics data reduction: it is the Lorentzian distribution. It is an appropriate distribution for describing data corresponding to resonance behavior, such as the variation with energy of the cross section of a nuclear reaction or the variation with velocity of the absorption of radiation in the Mössbauer effect.

The *Lorentzian distribution* function $P_L(x,\mu,\Gamma)$, also called the *Cauchy distribution*, is defined as

$$P_L(x,\mu,\Gamma) = \frac{1}{\pi} \frac{\Gamma/2}{(x - \mu)^2 + (\Gamma/2)^2} \tag{3-14}$$

This distribution is symmetric about its mean μ with a width characterized by its half-width Γ. The most striking difference between it and the Gaussian distribution is that it does not diminish to 0 as rapidly; the behavior for large deviations is proportional to the inverse of the square of the deviation, rather than exponentially related to the square of the deviation.

As with the Gaussian distribution, the Lorentzian distribution function is a continuous function, and the probability of observing a value x must be related to the interval within which the observations may fall. The probability $dP_L(x,\mu,\Gamma)$ for an observation to fall within an infinitesimal differential interval dx around x is given by the product of the probability function $P_L(x,\mu,\Gamma)$ and the size of the interval dx.

$$dP_L(x,\mu,\Gamma) = P_L(x,\mu,\Gamma)\ dx$$

The normalization of the probability function $P_L(x,\mu,\Gamma)$ is such that the integral of the probability over all possible values of x is unity

$$\int_{-\infty}^{\infty} P_L(x,\mu,\Gamma)\ dx = \frac{1}{\pi} \int_{-\infty}^{\infty} \frac{1}{1+z^2}\ dz = 1$$

where $z = (x - \mu)/(\Gamma/2)$.

Program 3-6 The calculation of Equation (3-14) is illustrated in the computer routine PLOREN of Program 3-6. This is a Fortran function subprogram to calculate the Lorentzian distribution function $P_L(x,\mu,\Gamma)$ for a given value of x, μ, and Γ. The input variables are X = x, AVERAG = μ, and WIDTH = Γ. The probability function $P_L(x,\mu,\Gamma)$ is evaluated directly in statement 1 and returned to the main program as the value of the function PLOREN.

Mean and standard deviation The mean μ of the Lorentzian distribution is given as one of the free parameters in Equation (3-14). It is obvious from the symmetry of the distribution about μ that μ must be equal to the mean as well as to the median and to the most probable value.

The standard deviation σ is not defined for the Lorentzian

Program 3-6 PLOREN Lorentzian probability distribution $P_L(x,\mu,\Gamma)$.

```
C      FUNCTION PLOREN
C
C      PURPOSE
C      EVALUATE LORENTZIAN PROBABILITY FUNCTION
C
C      USAGE
C        RESULT = PLOREN (X, AVERAG, WIDTH)
C
C      DESCRIPTION OF PARAMETERS
C        X       - VALUE FOR WHICH PROBABILITY IS TO BE EVALUATED
C        AVERAG  - MEAN OF DISTRIBUTION
C        WIDTH   - FULL WIDTH AT HALF MAXIMUM OF DISTRIBUTION
C
C      SUBROUTINES AND FUNCTION SUBPROGRAMS REQUIRED
C        NONE
C
C      MODIFICATIONS FOR FORTRAN II
C        NONE
C
       FUNCTION PLOREN (X, AVERAG, WIDTH)
     1 PLOREN = 0.1591549431 * WIDTH / ((X-AVERAG)**2 + (WIDTH/2.)**2)
       RETURN
       END
```

distribution as a consequence of its slowly decreasing behavior for large deviations. If we attempt to evaluate the expectation value for the square of the deviations,

$$\sigma^2 = \langle (x - \mu)^2 \rangle = \frac{1}{\pi} \frac{\Gamma^2}{4} \int_{-\infty}^{\infty} \frac{z^2}{1 + z^2}\, dz$$

we find that the integral is unbounded: the integral does not converge for large deviations.

The width of the Lorentzian distribution is instead characterized by the *full-width at half-maximum* Γ, generally called the *half-width*. This parameter is defined such that when the deviation from the mean is equal to one half the half-width, $x - \mu = \Gamma/2$, the probability function $P(x,\mu,\Gamma)$ is half its value at the maximum. Thus, the half-width Γ is the full-width of the curve measured at the level of half the maximum probability.

To show that this definition of Γ is consistent with its use in Equation (3-14), we note that at the peak of the distribution, the denominator is equal to the right-hand term alone. At the point where $x - \mu = \Gamma/2$, the left-hand term is equal to the right-hand term, and the denominator is twice as large as at the peak.

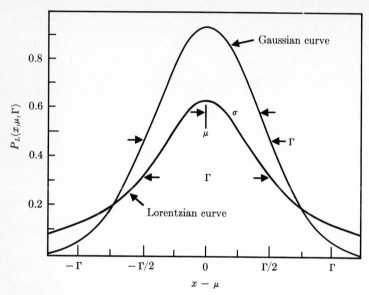

FIGURE 3-6 Lorentzian probability distribution $P_L(x,\mu,\Gamma)$ vs. $x - \mu$, compared with Gaussian function $P_G(x,\mu,\sigma)$ where $\sigma = \Gamma/2.345$. Higher narrower peak is Gaussian curve.

Illustration A comparison of the Lorentzian and the Gaussian distributions is given in Figure 3-6 with the distributions given the same values for the means μ and for the half-widths Γ. The standard deviation σ for the Gaussian distribution is determined from Equation (3-10). Both distributions are normalized according to their definitions of Equations (3-9) and (3-14).

In both cases the value of the maximum probability is inversely proportional to the half-width Γ and therefore dependent on the units of that parameter. For illustration, the half-width is specified as unity $\Gamma = 1$ for the curves of Figure 3-6. This results in a peak value of $2/\pi \simeq 0.64$ for the Lorentzian distribution and a peak value of $2.354/\sqrt{2\pi} \simeq 1.07$ for the Gaussian distribution, more than 50% larger. This difference in peak height graphically illustrates the fact that more of the contribution to the area is contained in the central peak for the Gaussian distribution than for the Lorentzian distribution with its significantly larger tail.

Except for the normalization, the Lorentzian distribution is equivalent to the dispersion relation which is used, for example, in describing the cross section of a nuclear reaction for a Breit-Wigner resonance.

$$\sigma = \pi\lambda^2 \frac{\Gamma_1\Gamma_2}{(E - E_0)^2 + (\Gamma/2)^2}$$

The important feature of the distribution is the behavior of the denominator.

SUMMARY

Binomial distribution: Describes the probability of observing x successes out of n tries when the probability for success in each try is p.

$$P_B(x,n,p) \equiv \binom{n}{x} p^x q^{n-x} = \frac{n!}{x!(n-x)!} p^x(1-p)^{n-x}$$

$$\mu = np \qquad \sigma^2 = np(1-p)$$

Poisson distribution: Limiting case of binomial distribution for large n and constant μ; appropriate for describing small samples of large populations.

$$P_P(x,\mu) \equiv \frac{\mu^x}{x!} e^{-\mu} \qquad \sigma^2 = \mu$$

Gaussian distribution: Limiting case of binomial distribution for large n and finite p; appropriate for smooth symmetric distributions.

$$P_G(x,\mu,\sigma) = \frac{1}{\sigma\sqrt{2\pi}} \exp\left[-\frac{1}{2}\left(\frac{x-\mu}{\sigma}\right)^2\right]$$

Half-width: $\qquad \Gamma = 2.354\sigma$
Probable error: $\quad$ P.E. $= 0.6745\sigma$

Lorentzian distribution: Dispersion relation

$$P_L(x,\mu,\Gamma) \equiv \frac{1}{\pi} \frac{\Gamma/2}{(x-\mu)^2 + (\Gamma/2)^2}$$

EXERCISES

3-1 Evaluate the following:
    (a) 3!   (b) 5!   (c) 8!/3!   (d) 22!/20!   (e) 10!

3-2 Evaluate the binomial distribution $P_B(x,n,p)$ for $n = 6$, $p = \frac{1}{2}$, and $x = 0 - 6$. Sketch the distribution and identify the mean and standard deviation.

3-3 Evaluate the following.

    (a) $\dbinom{6}{3}$ (b) $\dbinom{4}{2}$ (c) $\dbinom{10}{3}$ (d) $\dbinom{52}{4}$

3-4 Do Exercise 3-2 for $p = \frac{1}{6}$.

3-5 On a certain kind of dime slot machine there are 10 different symbols that can appear in each of three windows. The machine pays off different amounts when either one, two, or three lemons appear. How much should it pay for each of the three possibilities if it is to be honest?

3-6 Evaluate the Poisson distribution for Example 3-4.

3-7 Verify that if μ is an integer, the probability for the mean is equal to the probability for 1 less than the mean $P_P(\mu,\mu) = P_P(\mu - 1, \mu)$ for the Poisson distribution.

3-8 What is the standard deviation for a counting experiment with a mean $\mu = 100$ counts? What if the mean number of counts is increased by a factor of 4?

3-9 What is the probability $P_G(\mu + \sigma, \mu, \sigma)$ for the Gaussian distribution of observing a value 1 standard deviation from the mean? Compare it with the probability $P_G(\mu,\mu,\sigma)$ for the mean. Do the same for the P.E. and the half-width $\Gamma/2$.

3-10 What is the integral probability for the Gaussian distribution of observing a value within 1 standard deviation of the mean? Within the P.E.? Within the half-width $\Gamma/2$?

3-11 To a very good approximation, the entire area of a Gaussian curve is enclosed within $\mu \pm 3\sigma$. What fraction of the area of a Lorentzian curve is enclosed within $\mu \pm 3\Gamma/2$?

3-12 Show that the sum in Equation (3-4) reduces to $\mu = np$. Hint: Define $y = x - 1$ and $m = n - 1$ and use the fact that

$$\sum_{y=0}^{m} \left[\frac{m!}{y!(m-y)!} p^y (1-p)^{m-y} \right] = \sum_{y=0}^{m} P_B(y,m,p) = 1$$

3-13 Show that the sum in Equation (3-5) reduces to $\sigma^2 = np(1 - p)$. Hint: Use the results of Exercise 2-11 to simplify the expression. Define $y = x - 1$ and $m = n - 1$ and use the results of Exercise 3-12.

3-14 Show that the sum in Equation (3-8) reduces to $\sigma^2 = \mu$. Hint: Use the results of Exercise 2-11 to simplify the expression. Define $y = x - 1$ and show that the sum reduces to $\mu\langle y + 1 \rangle - \mu^2$. Use the results of Exercise 3-12 to show that $\langle y + 1 \rangle = \mu + 1$.

CHAPTER FOUR

PROPAGATION
OF ERRORS

4-1 GENERAL METHOD

In Chapter 2 we discussed methods for extracting from a set of data points estimates of the mean and standard deviation which describe, respectively, the desired result and the uncertainties of the data. In the next chapter we will be concerned with the effects these uncertainties will have on the determination of other parameters. In order to provide a framework for this, let us consider the way in which uncertainties are propagated or carried over from the data points to the parameters, and how the uncertainties of the determinations of some parameters of an experiment will be propagated to the final result.

EXAMPLE 4-1 Suppose we wish to find the volume V of a box with length L, width W, and height H. We can measure each of the three dimensions to be L_0, W_0, and H_0 and combine these measurements to yield a value for the volume.

$$V_0 = L_0 W_0 H_0$$

How do uncertainties in the estimates L_0, etc., affect the resulting uncertainty in the final result V_0?

If we knew the actual errors $\Delta L = L_0 - L$, etc., in each of the dimensions, we could estimate the error in the final result $\Delta V = V_0 - V$ by assuming it to be proportional to the errors in each of the dimensions. The error in V is approximately the sum of the products of the errors in each dimension times the effect that dimension has on the final value of V.

$$\Delta V \simeq \Delta L \left(\frac{\partial V}{\partial L}\right)_{W_0 H_0} + \Delta W \left(\frac{\partial V}{\partial W}\right)_{L_0 H_0} + \Delta H \left(\frac{\partial V}{\partial H}\right)_{L_0 W_0} \quad (4\text{-}1)$$

The terms in parentheses are the partial derivatives of V with respect to each of the dimensions. They are defined as the proportionality constants between changes in V and changes in the corresponding dimension evaluated at the point L_0, W_0, H_0 for infinitesimally small changes in the dimension. This approximation neglects higher-order terms in Taylor's expansion, which is equivalent to neglecting the fact that the partial derivatives are not constant over the range of L, W, and H given by their errors. If the errors are large, we must include in this definition at least second partial derivatives ($\partial^2 V / \partial L^2$) and partial cross derivatives ($\partial^2 V / \partial L\, \partial W$), but we will omit these from the discussion which follows.

For our example of $V = LWH$, Equation (4-1) reduces to

$$\Delta V \simeq W_0 H_0 \,\Delta L + L_0 H_0 \,\Delta W + L_0 W_0 \,\Delta H$$

which can be simplified by dividing through by $V_0 = L_0 W_0 H_0$.

$$\frac{\Delta V}{V_0} \simeq \frac{\Delta L}{L_0} + \frac{\Delta W}{W_0} + \frac{\Delta H}{H_0}$$

Uncertainties In general, however, we do not know the actual errors in the determinations of any of the parameters. What we may know, instead, is some characteristic of the uncertainty or estimated error of each parameter, such as the standard deviation σ of the distribution describing the probability of determining various values for that parameter. How can we combine the standard deviations of the individual parameters to estimate the uncertainty in the result?

Suppose we want to determine a quantity x which is a function of at least two other variables u and v which are actually measured. We will determine the characteristics of x from those for u and v and from the functional dependence.

$$x = f(u,v, \ . \ . \ .) \tag{4-2}$$

Although it may not always be exact, we will assume that the most probable value for x is given by

$$\bar{x} = f(\bar{u},\bar{v}, \ . \ . \ .) \tag{4-3}$$

The uncertainty in the resulting value for x can be found by considering the spread of the values of x resulting from combining the individual measurements $u_i, v_i, \ . \ . \ .$ into individual results x_i.

$$x_i = f(u_i, v_i, \ . \ . \ .)$$

In the limit of an infinite number of measurements, the mean of this distribution will coincide with the average $\bar{x}$ given in Equation (4-3) and we can use the definition of Equation (2-2) to find the variance σ_x^2 (which is the square of the standard deviation σ_x).

$$\sigma_x^2 = \lim_{N \to \infty} \frac{1}{N} \Sigma(x_i - \bar{x})^2 \tag{4-4}$$

Just as we expressed the deviations of V in Equation (4-1) as a function of the deviation of the dimensions L, W, and H, we can express the deviations $x_i - \bar{x}$ in terms of the deviations $u_i - \bar{u}$, $v_i - \bar{v}, \ . \ . \ .$ of the observed parameters

$$x_i - \bar{x} \simeq (u_i - \bar{u}) \left(\frac{\partial x}{\partial u}\right) + (v_i - \bar{v}) \left(\frac{\partial x}{\partial v}\right) + \ \cdot \ \cdot \ \cdot \tag{4-5}$$

where we have omitted specific notation of the fact that each of the partial derivatives is evaluated with all other variables fixed at their mean values.

Variance and covariance Combining Equations (4-4) and (4-5) we can express the variance σ_x^2 for x in terms of the variances σ_u^2, σ_v^2, . . . for the variables u, v, . . . which were actually measured.

$$\sigma_x^2 \simeq \lim_{N \to \infty} \frac{1}{N} \Sigma \left[(u_i - \bar{u}) \left(\frac{\partial x}{\partial u} \right) + (v_i - \bar{v}) \left(\frac{\partial x}{\partial v} \right) + \cdots \right]^2$$

$$\simeq \lim_{N \to \infty} \frac{1}{N} \Sigma \left[(u_i - \bar{u})^2 \left(\frac{\partial x}{\partial u} \right)^2 + (v_i - \bar{v})^2 \left(\frac{\partial x}{\partial v} \right)^2 \right.$$

$$\left. + 2(u_i - \bar{u})(v_i - \bar{v}) \left(\frac{\partial x}{\partial u} \right) \left(\frac{\partial x}{\partial v} \right) + \cdots \right] \quad (4\text{-}6)$$

The first two terms of Equation (4-6) can be expressed in terms of the variances σ_u^2 and σ_v^2 given by Equation (2-2).

$$\sigma_u^2 = \lim_{N \to \infty} \frac{1}{N} \Sigma (u_i - \bar{u})^2 \qquad \sigma_v^2 = \lim_{N \to \infty} \frac{1}{N} \Sigma (v_i - \bar{v})^2 \quad (4\text{-}7)$$

In order to express the third term of Equation (4-6) in a similar form, we introduce the *covariance* σ_{uv}^2 between the variables u and v defined analogous to the variances of Equations (4-7).

$$\sigma_{uv}^2 \equiv \lim_{N \to \infty} \frac{1}{N} \Sigma [(u_i - \bar{u})(v_i - \bar{v})]$$

With these definitions, the approximation for the standard deviation σ_x for x given in Equation (4-6) can be expressed as

$$\sigma_x^2 \simeq \sigma_u^2 \left(\frac{\partial x}{\partial u} \right)^2 + \sigma_v^2 \left(\frac{\partial x}{\partial v} \right)^2 + 2\sigma_{uv}^2 \left(\frac{\partial x}{\partial u} \right) \left(\frac{\partial x}{\partial v} \right) + \cdots$$

$$(4\text{-}8)$$

The first two terms are averages of squares of deviations, which will presumably dominate. If there are additional variables besides u and v in the determination of x, their contributions to the variance of x will have similar terms.

The third term is the average of cross terms involving

products of deviations in u and v simultaneously. If we can make the assumption that the fluctuations in u and v are uncorrelated, then on the average we would expect to find as many approximately equal negative values for this term as positive values, and we would expect the contribution to vanish in the limit of a large random selection of observations.

If the fluctuations in the observations of u and v, . . . are uncorrelated, Equation (4-8) reduces to

$$\sigma_x{}^2 \simeq \sigma_u{}^2 \left(\frac{\partial x}{\partial u}\right)^2 + \sigma_v{}^2 \left(\frac{\partial x}{\partial v}\right)^2 + \cdots \tag{4-9}$$

with additional similar terms corresponding to additional variables.

4-2 SPECIFIC FORMULAS

The expressions of Equations (4-8) and (4-9) were derived for the general relationship of Equation (4-2) giving x as an arbitrary function of u and v, ... In the following specific cases of functions $f(u,v, \ . \ . \ .)$, the parameters a and b are defined as positive constants and the parameters u and v are variables.

Addition and subtraction If x is the weighted sum of u and v,

$$x = au \pm bv$$

the partial derivatives are simply the weighting constants.

$$\left(\frac{\partial x}{\partial u}\right) = a \qquad \left(\frac{\partial x}{\partial v}\right) = \pm b$$

Equation (4-8) becomes

$$\sigma_x{}^2 = a^2\sigma_u{}^2 + b^2\sigma_u{}^2 \pm 2ab\sigma_{uv}{}^2 \tag{4-10}$$

Note that there is a possibility for the variance $\sigma_x{}^2$ to vanish if the covariance $\sigma_{uv}{}^2$ has the proper magnitude and sign. This would happen if the fluctuations were completely correlated so that each erroneous observation of u would be exactly compensated for by a corresponding erroneous observation of v.

For example, the perimeter P of a table is given as twice the sum of the length plus the width $P = 2L + 2W$. If the length L and width W have uncertainties given by

$$\sigma_L = 2 \text{ cm} \quad \text{and} \quad \sigma_W = 1 \text{ cm}$$

the uncertainty in the perimeter is given by

$$\sigma_P{}^2 = 4\sigma_L{}^2 + 4\sigma_W{}^2 = 20 \text{ cm}^2$$

or $\sigma_P = 4.5$ cm. If the length and width were each measured twice to evaluate the perimeter

$$P = L_1 + W_1 + L_2 + W_2$$

then the uncertainty in P would be less:

$$\sigma_P{}^2 = \sigma_{L_1}{}^2 + \sigma_{W_1}{}^2 + \sigma_{L_2}{}^2 + \sigma_{W_2}{}^2 = 2\sigma_L{}^2 + 2\sigma_W{}^2 = 10 \text{ cm}^2$$

or $\sigma_P = 3.2$ cm.

Multiplication and division If x is the weighted product of u and v,

$$x = \pm auv$$

the partial derivatives of each variable contain the values of the other variable.

$$\left(\frac{\partial x}{\partial u}\right) = \pm av \qquad \left(\frac{\partial x}{\partial v}\right) = \pm au$$

Equation (4-8) yields

$$\sigma_x{}^2 = a^2v^2\sigma_u{}^2 + a^2u^2\sigma_v{}^2 + 2a^2uv\sigma_{uv}{}^2$$

which can be expressed more symmetrically.

$$\frac{\sigma_x{}^2}{x^2} = \frac{\sigma_u{}^2}{u^2} + \frac{\sigma_v{}^2}{v^2} + 2\frac{\sigma_{uv}{}^2}{uv}$$

Similarly, if x is obtained through division

$$x = \pm \frac{au}{v}$$

the variance for x is given by

$$\frac{\sigma_x^2}{x^2} = \frac{\sigma_u^2}{u^2} + \frac{\sigma_v^2}{v^2} - 2 \frac{\sigma_{uv}^2}{uv} \tag{4-11}$$

For example, the area of a triangle is equal to half the product of the base times the height $A = \frac{1}{2}bh$. If the base and height have values of $b = 5$ cm and $h = 10$ cm and uncertainties given by $\sigma_b = 1$ mm and $\sigma_h = 3$ mm, the area is $A = 25$ cm² and the uncertainty in the area is given by

$$\sigma_A^2 \simeq 625 \left(\frac{\sigma_b^2}{25} + \frac{\sigma_h^2}{100} \right) = 81.25 \text{ mm}^2$$

or $\sigma_A = 9$ mm. Although the absolute uncertainty in the height is three times that of the base, the relative uncertainty σ_h/h is only $1\frac{1}{2}$ times as large and its contribution to the variance of the area is only $(1\frac{1}{2})^2$ times as large.

Powers If x is obtained by raising the variable u to a power

$$x = au^{\pm b}$$

the derivative of x with respect to u is

$$\left(\frac{\partial x}{\partial u} \right) = \pm abu^{\pm b-1} = \pm \frac{bx}{u}$$

and Equation (4-8) becomes

$$\frac{\sigma_x}{x} = b \frac{\sigma_u}{u}$$

For example, the area of a circle is proportional to the square of the radius $A = \pi r^2$. If the radius is determined to be $r = 10$ cm with an uncertainty $\sigma_r = 3$ mm, the area is $A = 100\pi$ cm² with an uncertainty given by $\sigma_A = 100\pi$ cm² $(2(0.3)$ cm$/10$ cm$) = 6\pi$ cm².

Exponentials   If x is obtained by raising the natural base to a power proportional to u,

$$x = ae^{\pm bu}$$

the derivative of x with respect to u is

$$\frac{\partial x}{\partial u} = \pm abe^{\pm bu} = \pm bx$$

and Equation (4-8) becomes

$$\frac{\sigma_x}{x} = b\sigma_u \tag{4-12}$$

If the constant which is raised to the power is not equal to e, it can be written in the same manner explicitly

$$x = a^{\pm bu} = (e^{\ln a})^{\pm bu} = e^{\pm(b \ln a)u}$$

(where ln indicates the natural logarithm) and solved in the same manner as above.

$$\frac{\sigma_x}{x} = (b \ln a)\sigma_u$$

For example, the activity of a radioactive source as a function of time is given by $A = A_0 e^{-t/\tau}$ where A_0 is the activity at the time $t = 0$, and τ is the natural lifetime, defined as the time when the activity is decreased by a factor of $1/e$. A source with an initial activity $A_0 = 1000$/sec and a lifetime of $\tau = 5$ days will have an activity after $t = 20$ days of

$$A = 1000/\text{sec } e^{-4} = 18.3/\text{sec}$$

If the uncertainty in the initial activity can be neglected, and the uncertainty in the time is $\sigma_t = 1$ hr $= 0.042$ days, the uncertainty in the final activity is $\sigma_A = 18.3/\text{sec } (0.042/5) = 0.15/\text{sec}$.

Logarithms   If x is obtained by taking the logarithm of u,

$$x = a \ln (\pm bu)$$

the derivative with respect to u is

$$\frac{\partial x}{\partial u} = \frac{a}{u}$$

and Equation (4-8) becomes

$$\sigma_x = a \frac{\sigma_u}{u}$$

which is essentially the inverse of Equation (4-12).

For example, given the source of the previous example, if we ask when the activity will be reduced to $A = 10$/sec with an uncertainty $\sigma_A = 1$/sec, the equation must be rewritten to give the time t explicitly $t = -\tau \ln (A/A_0) = 24$ days. The uncertainty in the result is given by

$$\sigma_t = \frac{-\tau \sigma_A}{A} = -5 \text{ days } (\tfrac{1}{10}) = -0.5 \text{ days}$$

SUMMARY

Covariance:

$$\sigma_{uv}^2 = \langle (u - \bar{u})(v - \bar{v}) \rangle.$$

Propagation of errors: Assume $x = f(u,v)$.

$$\sigma_x^2 = \sigma_u^2 \left(\frac{\partial x}{\partial u}\right)^2 + \sigma_v^2 \left(\frac{\partial x}{\partial v}\right)^2 + 2\sigma_{uv}^2 \left(\frac{\partial x}{\partial u}\right) \left(\frac{\partial x}{\partial v}\right)$$

For u and v uncorrelated, $\sigma_{uv}^2 = 0$.

Specific formulas:

$$x = au \pm bv: \quad \sigma_x^2 = a^2\sigma_u^2 + b^2\sigma_v^2 + 2ab\sigma_{uv}^2$$

$$x = \pm auv: \quad \frac{\sigma_x^2}{x^2} = \frac{\sigma_u^2}{u^2} + \frac{\sigma_v^2}{v^2} + 2\frac{\sigma_{uv}^2}{uv}$$

$$x = \pm \frac{au}{v}: \quad \frac{\sigma_x^2}{x^2} = \frac{\sigma_u^2}{u^2} + \frac{\sigma_v^2}{v^2} - 2\frac{\sigma_{uv}^2}{uv}$$

$$x = au^{\pm b}: \quad \frac{\sigma_x}{x} = b\frac{\sigma_u}{u}$$

$$x = ae^{\pm bu}: \quad \frac{\sigma_x}{x} = b\sigma_u$$

$$x = a \ln (\pm bu): \quad \sigma_x = a\frac{\sigma_u}{u}$$

EXERCISES

4-1 Find the relationship between the uncertainty σ_x in x as a function of the uncertainties σ_u and σ_v in u and v for the following functions:

(a) $x = \frac{1}{2}(u + v)$ (b) $x = \frac{1}{2}(u - v)$ (c) $x = 1/u^2$

(d) $x = uv^2$ (e) $x = u^2 + v^2$

4-2 If a table is round and its diameter is determined to within 1%, how well is its area known? Would it be better to determine its radius to within 1%?

4-3 The resistance R of a cylindrical conductor is proportional to its length L and inversely proportional to its cross-sectional area $A = \pi r^2$. Which should be determined with higher precision, r or L, to optimize the determination of R? How much higher?

4-4 The initial activity N_0 and the lifetime τ of a radioactive source are known with uncertainties of 1% each. For estimation of the activity $N = N_0 e^{-t/\tau}$, the error in the initial activity N_0 dominates for small t and vice-versa for large t (compared with τ). For what value of t/τ do the errors in N_0 and τ contribute equally to the uncertainty in N?

CHAPTER FIVE

ESTIMATES OF
MEAN AND ERRORS

In Chapter 2 we defined the mean μ of the parent distribution describing the probable dispersion of a random set of observations and noted that the most probable estimate of the mean μ is the average $\bar{x}$ of the observations. The justification of that statement is based on the assumption that the measurements are distributed according to the Gaussian distribution. In general, we expect the distribution of measurements to be either Gaussian or Poisson, but since these distributions are indistinguishable for most physical situations (see the discussion of the Poisson distribution in Section 3-2), we can assume the Gaussian distribution is obeyed.

Method of maximum likelihood   In our experiment we have observed a set of N data points which we assume to be a random selection from the infinite set of the parent population distributed according to the parent distribution. If the parent distribution is Gaussian with a mean μ and a standard deviation σ, then the probability dP_i for making any single observation x_i within an interval dx_i is given by

$$dP_i = \frac{1}{\sigma \sqrt{2\pi}} \exp\left[-\frac{1}{2}\left(\frac{x_i - \mu}{\sigma}\right)^2 \right] dx_i$$

For simplicity we will denote this by saying that the probability function P_i for making an observation x_i is given by

$$P_i = \frac{1}{\sigma \sqrt{2\pi}} \exp\left[-\frac{1}{2}\left(\frac{x_i - \mu}{\sigma}\right)^2 \right]$$

Since we do not, in general, know the mean μ of the distribution for a physical experiment, we must estimate it with some experimentally derived parameter. Let us call the estimate μ'. What formula for deriving μ' from the data will yield us the maximum likelihood that the parent distribution had a mean equal to μ?

If we hypothesize a trial parent distribution with a mean μ' and a standard deviation $\sigma' = \sigma$, the probability of observing the value x_i is given by

$$P_i(\mu') = \frac{1}{\sigma \sqrt{2\pi}} \exp\left[-\frac{1}{2}\left(\frac{x_i - \mu'}{\sigma}\right)^2 \right] \tag{5-1}$$

Considering the entire set of N observations, the probability for observing that set is given by the product of the individual $P_i(\mu')$

$$P(\mu') = \prod_{i=1}^{N} P_i(\mu') \tag{5-2}$$

where the symbol Π denotes the product of the N probabilities $P_i(\mu')$.

The product of the constants multiplying the exponential in

Equation (5-1) is the same as the constant to the Nth power, and the product of the exponentials is the same as the exponential of the sum of the arguments. Equation (5-2) reduces to

$$P(\mu') = \left(\frac{1}{\sigma\sqrt{2\pi}}\right)^N \exp\left[-\frac{1}{2}\sum\left(\frac{x_i - \mu'}{\sigma}\right)^2\right] \qquad (5\text{-}3)$$

According to the *method of maximum likelihood* (see any basic text in statistics), if we compare the probabilities $P(\mu')$ of obtaining our set of observations from various parent populations with different means μ' but with the same standard deviation $\sigma' = \sigma$, the probability is greatest that the data were derived from a population with $\mu' = \mu$; that is, the most likely population from which such a set of data might have come is assumed to be the correct one.

Calculation of the mean Using the method of maximum likelihood, the most probable value for μ' is the one which gives the maximum value for the probability $P(\mu')$ of Equation (5-3). Since this probability is the product of a constant times an exponential to a negative argument, maximizing the probability $P(\mu')$ is equivalent to minimizing the argument X of the exponential.

$$X = -\frac{1}{2}\sum\left(\frac{x_i - \mu'}{\sigma}\right)^2$$

Using the method of calculus described in Appendix A, the minimum value of a function X is one which yields a value of 0 for the derivative.

$$\frac{dX}{d\mu'} = -\frac{d}{d\mu'}\frac{1}{2}\sum\left(\frac{x_i - \mu'}{\sigma}\right)^2 = 0$$

In order to evaluate this derivative, we use the fact that the derivative of a sum is equal to the sum of the derivatives.

$$\frac{dX}{d\mu'} = -\frac{1}{2}\sum\frac{d}{d\mu'}\left(\frac{x_i - \mu'}{\sigma}\right)^2 = \sum\left(\frac{x_i - \mu'}{\sigma^2}\right) = 0 \qquad (5\text{-}4)$$

The most probable value for the mean is given by the last equality in Equation (5-4)

$$\Sigma(x_i - \mu') = 0 \qquad \text{or} \qquad \mu' = \bar{x} \equiv \frac{1}{N}\Sigma x_i \qquad (5\text{-}5)$$

which is the same as the definition given to $\bar{x}$ in Equation (2-7). This formula for estimating the mean maximizes the probability $P(\mu')$ of Equation (5-3) that the observed data came from a parent distribution with a mean $\mu' = \bar{x}$.

Weighting the data   In developing the probability $P(\mu')$ of Equation (5-3) from the individual probabilities $P_i(\mu')$ of Equation (5-1), we assumed that the data points were all extracted from the same parent population and that the same parent distribution was valid for all data points. In some circumstances, however, some data points might be measured with better or worse precision than others. We can express this quantitatively by assigning a different parent distribution to those data points with the same mean but with a correspondingly smaller or larger standard deviation σ_i.

If we assign to each data point x_i its own standard deviation σ_i representing the precision with which that particular data point was measured, Equation (5-3) giving the probability $P(\mu')$ that the observed set of N data points came from parent distributions with means $\mu_i = \mu'$ and standard deviations σ_i becomes

$$P(\mu') = \prod_{i=1}^{N} \left(\frac{1}{\sigma_i \sqrt{2\pi}}\right) \exp\left[-\frac{1}{2}\sum\left(\frac{x_i - \mu'}{\sigma_i}\right)^2\right]$$

Using the method of maximum likelihood, we must maximize this probability, which is equivalent to minimizing the argument in the exponential.

$$-\frac{1}{2}\frac{d}{d\mu'}\sum\left(\frac{x_i - \mu'}{\sigma_i}\right)^2 = \sum\left(\frac{x_i - \mu'}{\sigma_i^2}\right) = 0$$

The most probable value for the mean is therefore the weighted average of the data points

$$\mu' = \frac{\Sigma(x_i/\sigma_i{}^2)}{\Sigma(1/\sigma_i{}^2)} \tag{5-6}$$

where each data point x_i is weighted inversely by its own variance $\sigma_i{}^2$ in the sum.

Estimated error of the mean   What uncertainty σ_μ is associated with our determination of the mean μ' in Equation (5-6)? Each data point x_i was obtained with an uncertainty characterized by its standard deviation σ_i. Each of these data points contributes to the determination of the mean μ' and therefore each data point contributes some uncertainty to the determination of the final results. How are these contributions accumulated?

In Chapter 4 we developed a formula (see Equation (4-9)), for combining the uncertainties of terms contributing to a single result. The variance $\sigma_\mu{}^2$ of the mean μ' is equal to the sum of the variances $\sigma_i{}^2$ of the individual data points x_i weighted by the squares of the effects $\partial\mu'/\partial x_i$ which the data points have on the result.

$$\sigma_\mu{}^2 = \Sigma\left[\sigma_i{}^2\left(\frac{\partial\mu'}{\partial x_i}\right)^2\right] \tag{5-7}$$

This approximation neglects correlations between the measurements x_i as well as second- and higher-order terms in the expansion of the variance $\sigma_\mu{}^2$, but it should be a reasonable approximation so long as none of the data points contributes a major portion of the final result.

If the uncertainties of the data points are all equal $\sigma_i = \sigma$, we can use Equation (5-5) to find the effect $\partial\mu'/\partial x_i$ which each data point x_i has on the determination of the mean μ'.

$$\frac{\partial\mu'}{\partial x_i} = \frac{\partial}{\partial x_i}\left(\frac{1}{N}\Sigma x_i\right) = \frac{1}{N} \tag{5-8}$$

Combining Equations (5-7) and (5-8), the estimated error of the mean σ_μ is given by

$$\sigma_\mu{}^2 = \sum \left[\sigma^2 \left(\frac{1}{N} \right)^2 \right] = \frac{\sigma^2}{N} \tag{5-9}$$

If the uncertainties of the data points are not equal, we must evaluate $\partial \mu'/\partial x_i$ from the expression of Equation (5-6) for the mean μ'.

$$\frac{\partial \mu'}{\partial x_i} = \frac{\partial}{\partial x_i} \frac{\Sigma(x_i/\sigma_i{}^2)}{\Sigma(1/\sigma_i{}^2)} = \frac{1/\sigma_i{}^2}{\Sigma(1/\sigma_i{}^2)}$$

Substituting this result into Equation (5-7) yields a general formula for the uncertainty of the mean σ_μ.

$$\sigma_\mu{}^2 = \sum \frac{1/\sigma_i{}^2}{[\Sigma(1/\sigma_i{}^2)]^2} = \frac{1}{\Sigma(1/\sigma_i{}^2)} \tag{5-10}$$

5-2 INSTRUMENTAL UNCERTAINTIES

For many experiments the fluctuations in measurements are due to the finite precision of the instruments used in making the measurements and are not directly correlated with the values of the quantity measured. These fluctuations are symptomatic of the fact that the measurements are made with uncertainties; i.e., no random measurement is expected to be exactly accurate because the experimental equipment is not sufficiently precise to yield the correct value for each measurement. The uncertainty introduced into the final result is called an *instrumental uncertainty*, as opposed to statistical fluctuations to be described in the next section. The resulting errors in the final determination are classified as *instrumental errors*.

EXAMPLE 5-1 Consider the experiment of Example 2-1. The length of a table is measured 100 times with an uncertainty for each data point which is known to be $\sigma = 0.500$ cm. All of the measurements are made with the same experimental apparatus, and the finite precision of this apparatus, characterized by σ, results in a spread of observations over several centimeters grouped around the known length $x = 20.000$ cm. Figure 2-1 shows a comparison of the dispersion of the data points with the known and estimated parent distributions which are assumed to be Gaussian.

The length of the table is estimated, according to Equation (2-7), to be $\mu \simeq \bar{x} = 20.028$ cm, and the precision of the experimental points is estimated, according to Equation (2-10), to be $\sigma \simeq s = 0.48$ cm. Since the uncertainties of all the data points are equal $\sigma_i = \sigma$, the arguments leading to Equation (5-9) are valid and we can estimate the uncertainty in our determination of the mean to be $\sigma_\mu \simeq s/\sqrt{N}$ where $N = 100$. This yields an uncertainty of $\sigma_\mu \simeq 0.05$ cm which is only slightly larger than the actual error of 0.028 cm.

Equal uncertainties For experiments such as that of Example 5-1 where the uncertainties are all equal $\sigma_i = \sigma$, Equation (5-5) gives the best estimate of the mean μ.

$$\mu \simeq \bar{x} = \frac{1}{N} \Sigma x_i \tag{5-11}$$

The uncertainty of the data points σ can be estimated from the data, if it is not known, according to Equation (2-10).

$$\sigma \simeq s = \sqrt{\frac{1}{N-1} \Sigma (x_i - \bar{x})^2} \tag{5-12}$$

Combining Equations (5-9) and (5-12), we can estimate the uncertainty σ_μ in the determination of the mean μ

$$\sigma_\mu = \frac{\sigma}{\sqrt{N}} \simeq \frac{s}{\sqrt{N}} \tag{5-13}$$

By similar analysis, we could estimate the uncertainty of σ_μ itself as well.

Unequal uncertainties If the uncertainties of some data points are different from those of others because of changes in the experimental conditions, the mean μ must be estimated according to Equation (5-6)

$$\mu \simeq \frac{\Sigma(x_i/\sigma_i{}^2)}{\Sigma(1/\sigma_i{}^2)} \tag{5-14}$$

and the uncertainty in the mean μ is estimated according to Equation (5-10)

$$\sigma_\mu{}^2 \simeq \frac{1}{\Sigma(1/\sigma_i{}^2)} \tag{5-15}$$

where the σ_i must be known from other experiments.

It often happens that the relative values of the σ_i are known, but the absolute magnitudes are not. For example, if one set of data are acquired with one scale range and another set with a different scale range, the σ_i may be equal within each set but differ by a known factor between the two sets, as would be the case if σ_i is proportional to the scale range. In such a case, the *relative* values of $\sigma_i' \simeq \sigma_i$ should be included as weighting factors in the determination of the mean μ and its uncertainty, and the *absolute* magnitudes of the σ_i can be estimated from the dispersion of the data points around the mean.

Since Equation (5-14) is independent of the absolute magnitudes of the σ_i, its form is unchanged.

$$\mu \simeq \frac{\Sigma(w_i' x_i)}{\Sigma w_i'} \qquad w_i' \equiv \frac{1}{\sigma_i'^2}$$

The absolute magnitudes of the σ_i can be estimated from the *average variance* of the data σ^2, evaluated as in Equation (5-12).

$$w_i \equiv \frac{1}{\sigma_i{}^2} = \frac{1}{\sigma^2}\frac{w_i'}{\Sigma w_i'/N} \qquad \sigma_i{}^2 = \sigma^2\frac{\sigma_i'^2}{N}\sum\frac{1}{\sigma_i'^2}$$

$$\sigma^2 \equiv \frac{1}{w} = \frac{N}{\Sigma w_i} \simeq s^2 = \frac{N\Sigma[w_i'(x_i - \bar{x})^2]}{(N-1)\Sigma w_i'}$$

The uncertainty in the mean σ_μ follows from Equation (5-13) or (5-14).

$$\sigma_\mu{}^2 = \frac{1}{\Sigma(1/\sigma_i'^2)} = \frac{\sigma^2}{N}$$

EXAMPLE 5-2   An experiment is performed to determine the voltage of a standard cell. The experimenter makes 40 measurements with the apparatus and finds a result $\bar{x}_1 = 1.022$ V with a spread of $s_1 = 0.01$ V in the observations. After the instructor shows him how to improve the equipment to decrease the uncertainty by a factor of 2.5 ($s_2 = 0.004$ V), the student makes 10 more measurements which yield a result $\bar{x}_2 = 1.018$ V.

The mean of all these observations is given by Equation (5-14).

$$\mu \simeq \frac{\dfrac{40(1.022) \text{ V}}{.01^2} + \dfrac{10(1.018) \text{ V}}{.004^2}}{\dfrac{40}{.01^2} + \dfrac{10}{.004^2}}$$

$$= .39(1.022) \text{ V} + .61(1.018) \text{ V} = 1.0196 \text{ V}$$

The uncertainty in the mean σ_μ is given by Equation (5-15).

$$\sigma_\mu \simeq \left(\frac{40}{.01^2} + \frac{10}{.004^2} \right)^{-\frac{1}{2}} = 0.0010 \text{ V}$$

Even though there were four times as many observations made with the higher uncertainty, the mean is nearly twice as dependent on the result of the second half of the experiment where the uncertainty of the measurements is smaller by a factor of 2.5.

The precision of the final result is better than that for either part of the experiment. The uncertainties in the estimates of the means μ_1 and μ_2 determined from the two sets of data independently are given by Equation (5-13).

$$\sigma_{\mu_1} = \frac{0.01 \text{ V}}{\sqrt{40}} = 0.0016 \text{ V} \qquad \sigma_{\mu_2} = \frac{0.004 \text{ V}}{\sqrt{10}} = 0.0013 \text{ V}$$

A comparison of these values illustrates the fact that taking more measurements decreases the resulting uncertainty only by the square root of the number of observations, which for this case is not so important as decreasing σ_i.

Program 5-1   The method of calculation of the mean and uncertainties is illustrated with the computer routine XFIT of

Program 5-1. This is a Fortran subroutine to calculate the mean μ and the standard deviations of the data σ and the mean σ_μ for a set of N data points. The input variables are X, SIGMAX, NPTS, and MODE, and the output variables are XMEAN, SIGMAM, and SIGMA.

The data points x_i are assumed to be stored in the array X with $X(I) = x_i$ for $I = i$ ranging from 1 to N. The standard deviations σ_i associated with the data points are assumed to be stored in the array SIGMAX with the ordering identical for the two arrays. The variable NPTS = N represents the number of data points.

The variable MODE determines whether the uncertainties in the data points are supposed to be supplied by the calling program or calculated from the data. If MODE = $+1$ (or any other positive nonzero integer), the subroutine uses the standard deviations σ_i stored in SIGMAX to calculate the weighting factor WEIGHT = 1./ SIGMAX(I)**2 in statement 24. If MODE = 0, the calculation proceeds on the assumption that the uncertainties of the data points are all equal $\sigma_i = \sigma$, and the weighting factor is ignored (set = 1. in statement 22). In this case, the array SIGMAX is totally ignored by the subroutine and is neither used nor altered. To save space in the main calling program, it is possible to omit the array SIGMAX and to substitute for it in the calling sequence the name of any other valid array, such as the data array X.

The numerator (SUMX) and denominator (SUM) of Equation (5-14) are accumlated in statements 31 and 32 as part of the sum over i from 1 to N in statements 20–32. The mean XMEAN = μ is evaluated in statement 41 according to Equation (5-14). If MODE = 0, indicating equal uncertainties $\sigma_i = \sigma$, the evaluation of the mean according to Equation (5-14) is identical to that according to Equation (5-11) since WEIGHT = 1. means that SUMX = Σx_i and SUM = N.

The spread of the data points SIGMA = σ is calculated in statements 51–54 according to the estimate of Equation (5-12). The uncertainty in the determination of the mean SIGMAM = σ_μ is evaluated in statement 66 according to Equation (5-15) for MODE > 0 or in statement 64 according to Equation (5-13) for MODE = 0 (with SUM = N).

In statement 70, the subroutine returns to the main program

Program 5-1 XFIT   Least-squares fit for a single variable.

```
C         SUBROUTINE XFIT
C
C         PURPOSE
C           CALCULATE THE MEAN AND ESTIMATED ERRORS FOR A SET OF DATA POINTS
C
C         USAGE
C           CALL XFIT (X, SIGMAX, NPTS, MODE, XMEAN, SIGMAM, SIGMA)
C
C         DESCRIPTION OF PARAMETERS
C           X      - ARRAY OF DATA POINTS
C           SIGMAX - ARRAY OF STANDARD DEVIATIONS FOR DATA POINTS
C           NPTS   - NUMBER OF DATA POINTS
C           MODE   - DETERMINES METHOD OF WEIGHTING
C                    +1 (INSTRUMENTAL) WEIGHT(I) = 1./SIGMAX(I)**2
C                     0 (NO WEIGHTING) WEIGHT(I) = 1.
C                    -1 (STATISTICAL)  WEIGHT(I) = 1.
C           XMEAN  - WEIGHTED MEAN
C           SIGMAM - STANDARD DEVIATION OF MEAN
C           SIGMA  - STANDARD DEVIATION OF DATA
C
C         SUBROUTINES AND FUNCTION SUBPROGRAMS REQUIRED
C           NONE
C
C         MODIFICATIONS FOR FORTRAN II
C           OMIT DOUBLE PRECISION SPECIFICATIONS
C           CHANGE DSQRT TO SQRTF IN STATEMENTS 54, 62, 64, AND 66
C
          SUBROUTINE XFIT (X, SIGMAX, NPTS, MODE, XMEAN, SIGMAM, SIGMA)
          DOUBLE PRECISION SUM, SUMX, WEIGHT, FREE
          DIMENSION X(1), SIGMAX(1)
C
C             ACCUMULATE WEIGHTED SUMS
C
       11 SUM = 0.
          SUMX = 0.
          SIGMA = 0.
          SIGMAM = 0.
       20 DO 32 I=1, NPTS
       21 IF (MODE) 22, 22, 24
       22 WEIGHT = 1.
          GO TO 31
       24 WEIGHT = 1. / SIGMAX(I)**2
       31 SUM = SUM + WEIGHT
       32 SUMX = SUMX + WEIGHT*X(I)
C
C             EVALUATE MEAN AND STANDARD DEVIATIONS
C
       41 XMEAN = SUMX/SUM
       51 DO 52 I=1, NPTS
       52 SIGMA = SIGMA + (X(I)-XMEAN)**2
          FREE = NPTS-1
       54 SIGMA = DSQRT(SIGMA/FREE)
       61 IF (MODE) 62, 64, 66
       62 SIGMAM = DSQRT(XMEAN/SUM)
          GO TO 70
       64 SIGMAM = SIGMA / DSQRT(SUM)
          GO TO 70
       66 SIGMAM = DSQRT(1./SUM)
       70 RETURN
          END
```

leaving XMEAN $= \mu$, SIGMAM $= \sigma_\mu$, and SIGMA $= \sigma$ as arguments of the calling sequence for output.

5-3 STATISTICAL FLUCTUATIONS

For many experiments it is more accurate to determine the standard deviations σ_i from a knowledge of the estimated parent distribution than to evaluate them from the data or from other experiments. If the observations are known to follow the Gaussian distribution, the standard deviation σ is a free parameter and must be determined experimentally. If the observations are known to be distributed according to the Poisson distribution, however, the standard deviation can be determined directly from a knowledge of the mean according to Equation (3-6).

As discussed in Section 3-2, the Poisson distribution is appropriate for describing the spread of data points in counting experiments where the numbers of items detected per unit time interval constitute the observations. For such experiments, the measurements fluctuate from observation to observation not because of any imprecision in measuring the time interval or because of any inexactness in counting the number of events occurring the interval, but because random samples of events distributed randomly in time contain numbers of events which fluctuate from sample to sample.

In any given time interval there is a finite nonzero chance of observing any integral number of events. The probability for observing any specific number of counts is given by the Poisson probability function, with a specified mean μ and a resulting uncertainty $\sigma = \sqrt{\mu}$. Since the fluctuations in the observations result from the statistical nature of the distribution, they are classified as *statistical fluctuations*, and the resulting errors in the final determination are classified as *statistical errors*.

Mean and standard deviation For reasonable values of the mean $\mu \gg 1$, the shape of the Poisson distribution is closely approximated by the Gaussian distribution. Therefore, we can use the same formula for estimating the mean developed in

Equation (5-11) of Section 5-2, assuming that all of the data points were extracted from the same parent population and therefore have the same uncertainties $\sigma_i = \sigma$.

$$\mu \simeq \bar{x} = \frac{1}{N}\, \Sigma x_i$$

According to Equation (3-7), the variance σ^2 for a Poisson distribution is equal to the mean μ.

$$\sigma^2 = \mu \simeq \bar{x} \tag{5-16}$$

The uncertainty in the mean σ_μ is obtained by combining Equations (5-9) and (5-16).

$$\sigma_\mu = \frac{\sigma}{\sqrt{N}} \simeq \sqrt{\frac{\bar{x}}{N}} \tag{5-17}$$

In some experiments, as in Example 5-2, part of the data might be obtained with a different uncertainty σ_i from that of the rest of the data. For statistical fluctuations, this implies that the data were obtained for different lengths of the (time) intervals. Since the mean number of events per interval μ is inversely proportional to the length of that interval, the uncertainty $\sigma = \sqrt{\mu}$ associated with the data is inversely proportional to the square root of the length of the interval.

If the uncertainties σ_i are not equal, it is necessary to use Equation (5-14) to evaluate the mean.

$$\mu \simeq \frac{\Sigma(x_i/\sigma_i{}^2)}{\Sigma(1/\sigma_i{}^2)}$$

The data x_i are evaluated by normalizing the actual measurements x_i' to the lengths of the intervals (for example, time intervals Δt_i),

$$x_i = \frac{x_i'}{\Delta t_i}$$

but the uncertainties σ_i associated with the data must be determined from the actual measurements x_i'

$$\sigma_i \simeq \sqrt{\bar{x}_i'} \tag{5-18}$$

where $\bar{x}_i'$ is the average of all the measurements that are made with the same uncertainty σ_i as the measurement x_i'.

Sample calculation The subroutine XFIT of Program 5-1 illustrates the calculations of Equations (5-11), (5-16), and (5-17). If the variable MODE $= -1$ (or any other negative integer), the mean XMEAN $= \mu$ and the uncertainty in the data SIGMA $= \sigma$ are calculated as for instrumental uncertainties with $\sigma_i = \sigma$. Thus, the standard deviation of the data σ represents the actual spread of the data points $\sigma = \langle (x_i - \mu)^2 \rangle$ rather than the theoretical estimate of Equation (5-16). The uncertainty in the mean σ_μ is evaluated in statement 62 according to Equation (5-17) (with SUM $= N$).

For data with unequal uncertainties, the variable MODE should be set equal to $+1$ and the uncertainties σ_i of the data points as given in Equation (5-18) stored in the array SIGMAX. The calculation then proceeds as for instrumental uncertainties according to Equations (5-14) and (5-15).

EXAMPLE 5-3 The activity of a radioactive source is measured $N = 10$ times with a time interval $\Delta t = 1$ min. The data are given in Table 5-1. The average of these data points is $\bar{x} = 20.1$ counts/min calculated according to Equation (5-11). The spread of the data points is characterized by $\sigma \simeq 4.5$ counts/min calculated according to Equation (5-16). The uncertainty in the mean is calculated according to Equation (5-17) to be $\sigma_{\bar{x}} \simeq 1.4$ counts/min.

If we were to combine the data into one observation $x' = \Sigma x_i$ for one 10-min interval, we would obtain the same result. The activity is $x' = 201$ counts/10 min $= 20.1$ counts/min as before. The uncertainty in the result is given by the standard deviation of the single data point $\sigma_{x'} = \sigma' \simeq 14$ counts/10 min $= 1.4$ counts/min.

If we were to make an additional measurement $x_{11} = 220$ counts for a 10-min period, we could combine x' and x_{11} exactly as above to obtain a total $x_T = x' + x_{11} = 21$ counts/min with an uncertainty $\sigma_{\mu T} = \sqrt{x_T} = 1$ count/min which is smaller

Table 5-1 Experimental data for the activity of a radioactive source from the experiment of Example 5-3. The data tabulated are the number of counts x_i detected in each time interval Δt_i

Interval t_i, min	Counts x_i
1	14
1	28
1	27
1	16
1	13
1	14
1	30
1	15
1	27
1	17
10	220
20	421

$$\bar{x} = \frac{1}{N}\Sigma x_i = 201 \text{ counts}/10 \text{ min} = 20.1 \text{ counts/min}$$

$$\sigma \simeq \sqrt{\bar{x}} = 4.5 \text{ counts}/1 \text{ min} = 4.5 \text{ counts/min}$$

$$\sigma_{\bar{x}} \simeq \sqrt{\frac{\bar{x}}{N}} = \sqrt{2.01} = 1.4 \text{ counts/min}$$

$$\sigma_{11} \simeq \sqrt{220} = 15 \text{ counts}/10 \text{ min} = 1.5 \text{ counts/min}$$

$x_T = 201 + 220 = 421 \text{ counts}/20 \text{ min} = 21.0 \text{ counts/min}$

$\sigma_{\mu T} \simeq \sqrt{x_T} = 20.5 \text{ counts}/20 \text{ min} = 1.0 \text{ counts/min}$

$\bar{x}_T = \dfrac{\Sigma(x_i/\sigma_i^2)}{\Sigma(1/\sigma_i^2)} = \dfrac{{}^{201}\!/_{20.1} + {}^{22}\!/_{2.2}}{{}^{10}\!/_{20.1} + {}^{1}\!/_{2.2}} \text{ counts/min}$

$\quad = 21.0 \text{ counts/min}$

$\sigma_T = \left(\dfrac{10}{\sigma^2} + \dfrac{1}{\sigma_{11}^2}\right)^{-\frac{1}{2}} \simeq \left(\dfrac{10}{20.1} + \dfrac{1}{2.2}\right)^{-\frac{1}{2}} = 1.0 \text{ count/min}$

than $\sigma_{\bar{x}}$ by a factor of $\sqrt{2}$. We could, alternatively, combine the original data points x_i with the new data point x_{11} to evaluate the average $\bar{x}_T$ according to Equation (5-14). The uncertainty in the final result σ_T comes from combining the uncertainties of the individual data points according to Equations (5-15) and (5-18).

Note that we could have simplified matters greatly by considering all the data as one experimental point: $x = 421 \text{ counts}/20 \text{ min} = 21 \text{ counts/min}$ with an uncertainty given by the square root of the total number of counts $\sigma_x \simeq \sqrt{421} = 20.5 \text{ counts}/20 \text{ min} = 1.0 \text{ count/min}$.

5-4 χ^2 TEST OF DISTRIBUTION

Once we have calculated the mean and standard deviation from our data, we may be in a position to say even more about the parent population. If we can be fairly confident of the type of parent distribution which describes the spread of the data points (e.g., Gaussian or Poisson distribution), then we can describe the parent distribution in detail and predict the outcome of future experiments from a statistical point of view. But how can we be sure of the shape of the parent distribution? What test is there which will justify or contradict our choice of distribution?

Since we are concerned with the behavior of the probability function $P(x_i)$ as a function of the observed values x_i, a complete discussion will be postponed until Chapter 10 following the development of procedures to compare data with complex functions. Let us for the moment use the results of Section 10-1 without derivation. The test which we will describe here is the χ^2 test for goodness of fit.

Probability distribution If measurements x_i are made of the quantity x, we can group the observations into frequencies of identical observations. In Table 2-1, for example, the data for measurements of the length of a table are given only to the nearest millimeter and are listed as a frequency f of observations of each value for each of the 24 different values observed. Let us call the frequency or number of observations $f_1(x_j)$ for each different measured value x_j. We denote the number of different measured values n so that j runs from 1 to n. If the probability for observing the value x_j in any random measurement is denoted $P(x_j)$, the expected number of such observations is $f(x_j) = NP(x_j)$ where N is the total number of measurements.

For any measured value x_j, there is a standard deviation $\sigma_j(f)$ associated with the uncertainty in determining the frequency $f(x_j)$. This is not the same as the uncertainty σ_i associated with the deviation $x_i - \mu$ between the value of an individual measurement x_i and the mean μ. If we were to set out to determine the distribution function $P(x_i)$ very precisely by taking a number of sets of

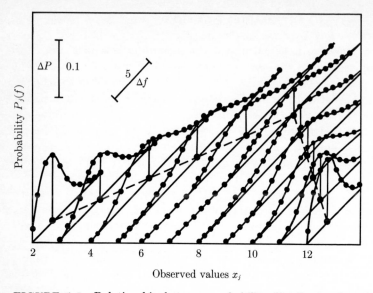

FIGURE 5-1 Relationship between probability distribution function $P(x_i)$ (dashed line), which determines the probability of observing x_i in any random measurement i, vs. the functions $P_j(f_k)$ (solid curves), which determine the probabilities of observing frequencies $f_k(x_j)$ for the possible values x_j in any random set of measurements k. Large dots indicate the expected frequencies $\langle f(x_j) \rangle = NP(x_j)$; small dots indicate the possible frequencies $f_k(x_j)$.

data ($k = 1, n_k$), then we would have several measurements $f_k(x_j)$ for each frequency $f(x_j)$. The spread in these measurements, for various values of k, is characterized by $\sigma_j(f)$.

EXAMPLE 5-4 Figure 5-1 illustrates the different meanings for σ_i and $\sigma_j(f)$. If we toss two dice 108 times and observe the sums x_i of the values showing on the two dice, the possible values of the sum x_j are the integers from 2 to 12 ($n = 11$). The probability function $P(x_j)$ which predicts the frequency $f(x_j) = NP(x_j)$ of observing each value x_j is given in Table 5-2 for $j = 1$ to 6 ($x_j = 2$ to 7). For higher values of j, the distribution is symmetric: $f(x_j) = f(x_{n-j+1})$ where $n = 11$. This distribution is shown in an isometric view as the dashed curve of Figure 5-1.

Table 5-2 Probability functions of Example 5-4 pertaining to the sum of the points showing on two dice. $P(x_i)$ is the parent distribution function which determines the probability of observing the sum x_i in any random measurement. The probability functions $P_j(f_k)$ are the parent distribution functions pertaining to each observed sum x_j which determine the probability of observing the frequency $f_k(x_j)$ instead of the expected frequency $f(x_j)$ in any random set of measurements. The total number of measurements is $N = 108$

j.........	1	2	3	4	5	6
x_j........	2	3	4	5	6	7
$NP(x_j)$...	3	6	9	12	15	18
k			$P_k(x_j)$			
0	.050	.002	.000	.000	.000	.000
1	.150	.015	.001	.000	.000	.000
2	.224	.045	.005	.000	.000	.000
3	.224	.089	.015	.002	.000	.000
4	.168	.134	.034	.005	.001	.000
5	.101	.161	.061	.012	.002	.000
6	.050	.161	.091	.025	.005	.001
8	.009	.103	.132	.065	.019	.004
10	.001	.041	.118	.105	.049	.015
12		.011	.075	.114	.081	.037
14		.002	.033	.090	.100	.066
16			.011	.054	.094	.089
18			.003	.025	.069	.094
20				.010	.041	.081
22				.003	.020	.060
24				.001	.008	.034
26					.003	.016
28					.001	.007
30						.003

For each value x_j, there is a probability distribution curve $P_j(f_k)$ which describes the distribution of possible determinations $f_k(x_j)$ of the frequency $f(x_j)$ with which the measured value x_j is observed. These distributions are indicated as solid curves in

Figure 5-1, graphed as the probability function $P_j(f_k)$ vs. the possible values of observed number $f_k(x_j)$.

The distribution of frequencies $P(x_j)$ is triangular, but the probability functions for the spread of measurements of each frequency $P_j(f_k)$ are Poisson distributions as illustrated in Figure 5-1. These functions are tabulated in Table 5-2 for each value of x_j and representative values of the frequency $f_k(x_j)$.

Definition of χ^2 With the definitions given above for n, N, x_j, $f_k(x_j)$, $P(x_j)$ and $\sigma_j(f)$, the definition of χ^2 from Section 10-1 is

$$\chi^2 \equiv \sum_{j=1}^{n} \frac{[f_1(x_j) - NP(x_j)]^2}{\sigma_j(f)^2} \tag{5-19}$$

In most experiments, however, we do not know what the values of $\sigma_j(f)$ are since we make only one set of measurements ($k = 1$). Fortunately, these uncertainties can be estimated from the data directly without measuring them explicitly.

If we consider how the measurements of $f_1(x_j)$ are made, we see that they should be distributed according to the Poisson distribution as shown in Figure 5-1. There exists a parent population for the measurements x_i distributed according to the parent distribution. For each value x_j, we have extracted a proportionate random sample of the parent population for that value. The fluctuations in the observed frequencies $f(x_j)$ come from the statistical probabilities of making random selections of finite numbers of items.

For the Poisson distribution, the variance $\sigma_j(f)^2$ is known to be equal to the mean $f(x_j) = NP(x_j)$ of the distribution. Therefore, we can simplify Equation (5-19) for such cases.

$$\chi^2 = \sum_{j=1}^{n} \frac{[f_1(x_j) - NP(x_j)]^2}{NP(x_j)} \tag{5-20}$$

Test of χ^2 As defined in Equations (5-19) and (5-20), χ^2 is a statistic which characterizes the dispersion of the observed frequencies from the expected frequencies. The numerator of

Equation (5-19) is a measure of the spread of the observations; the denominator is a measure of the expected spread. In order to test the goodness of fit of the observed frequencies to the assumed probability distribution, we must know how χ^2 is distributed; i.e., we need to know the probability of observing our calculated value of χ^2 from a random sample of data. If such a value is highly probable, then we can have confidence in our assumed distribution and vice versa.

If the observed frequencies agree exactly with the predicted frequencies $f_1(x_j) = f(x_j) = NP(x_j)$, then $\chi^2 = 0$. For any physical experiment where the predicted and observed frequencies will not be equal, we would expect a value of $\chi^2 \simeq n$, indicating that the observed and predicted spreads in the observations were, on the average, approximately equal for each value x_j. Larger values of χ^2 indicate larger deviations than expected from the assumed distribution. Rather than consider the probability of obtaining any particular value of χ^2, we will use an integral test to determine the probability of observing such a large value of χ^2.

Table C-4 gives values of the reduced chi-square $\chi_\nu^2 = \chi^2/\nu$ for different values of the number of degrees of freedom ν and as a function of the probability that any random sample of data points will have a value of χ^2 at least as large. The number of degrees of freedom is the number n of values of x_j used to calculate χ^2 minus the number of parameters calculated from the data to describe the distribution. For a Gaussian distribution, the data are used to calculate the mean and standard deviation of the probability distribution. The height is determined from these parameters and from the total number N of measurements. Thus, the number of degrees of freedom is given by

$$\nu = n - 2$$

For a given experiment with ν degrees of freedom and an observed value of χ^2, Table C-4 gives the probability that a random sample of data points would yield a value of χ^2 as large as or larger than that observed if the parent distribution were equal to the assumed distribution. If the probability is nearly equal to 1, the assumed distribution describes the spread of data points well.

If the probability is small, either the assumed distribution is not a good estimate of the parent distribution or the data are not a representative sample. There is no yes- or no-answer to the test; in fact, we would expect to find a probability of 0.5 with $\chi^2/\nu \simeq 1$, since statistically the observed values of χ^2 should exceed the norm half the time. But for most cases, the probability is either reasonably large or unreasonably small, and the test is fairly conclusive.

Sample calculation For example, in the experiment of Example 2-1, suppose we were to test how well the data agreed with the experimentally determined probability distribution. The observations x_j and the frequencies $f(x_j)$ with which these measurements were observed are reproduced in Table 5-3. The parent distribution $P(x_j)$ for these data is a Gaussian distribution with a mean $\mu = 20.000$ cm and a standard deviation $\sigma = 0.50$ cm. This is the distribution illustrated with the solid curve of Figure 2-1.

In Table 5-3, the dimensionless deviations of the measured values $|x_j - \mu|/\sigma$ are given in the third column. The expected distribution ϕ_0 is listed in the fourth column calculated from the tabulated values of $P_G(x,\mu,\sigma)$ for the Gaussian distribution given in Table C-1.

$$\phi_0 = N \, \Delta z P_G(z,0,1) \qquad z = \frac{|x - \mu|}{\sigma}, \; \Delta z = \frac{\Delta x}{\sigma}$$

The distribution ϕ_0 is normalized to the total number of data points $N = 100$ and to the separation of the data points $\Delta z = 0.2$.

$$\phi_0 = 20 P_G(z,0,1)$$

The contributions to χ^2 are calculated according to Equation (5-20) and are listed with their sum in column five of Table 5-3. With 24 different values of x_j, the number of degrees of freedom for fitting to the parent distribution is $\nu_0 = 24$ since none of the parameters of the parent distribution are determined from the data. The value of the reduced chi-square is $\chi^2_{\nu_0} = 0.87$, which corresponds to a probability of about 0.6 from Table C-4 that such a value would be exceeded in a random set of data. Since we

Table 5-3 χ^2 analysis of the data from Example 2-1 measuring the length of a block. Parameters for the parent Gaussian distribution ϕ_0 are $\mu = 20.000$ cm and $\sigma = 0.50$ cm; parameters for the estimated distribution ϕ are $\bar{x} = 20.028$ cm and $s = 0.48$ cm

Length	Fre-quency	$\dfrac{|x - \mu|}{\sigma}$	ϕ_0	$\dfrac{(f - \phi_0)^2}{\phi_0}$	$\dfrac{|x - \bar{x}|}{s}$	ϕ	$\dfrac{(f - \phi)^2}{\phi}$
18.9	1	2.2	0.71	0.12	2.35	0.52	0.44
19.0	0	2.0	1.08	1.08	2.14	0.84	0.84
19.1	1	1.8	1.58	0.21	1.92	1.29	0.07
19.2	2	1.6	2.22	0.02	1.73	1.86	0.01
19.3	1	1.4	2.99	1.32	1.52	2.61	0.99
19.4	4	1.2	3.88	0.00	1.31	3.52	0.07
19.5	3	1.0	4.84	0.70	1.10	4.55	0.53
19.6	9	0.8	5.79	1.78	0.89	5.60	2.06
19.7	8	0.6	6.66	0.27	0.68	6.60	0.30
19.8	11	0.4	7.37	1.79	0.48	7.41	1.73
19.9	9	0.2	7.82	0.18	0.27	8.00	0.12
20.0	5	0.0	7.98	1.11	0.06	8.30	1.31
20.1	7	0.2	7.82	0.09	0.15	8.21	0.18
20.2	8	0.4	7.37	0.05	0.36	7.80	0.01
20.3	9	0.6	6.66	0.82	0.58	7.02	0.56
20.4	6	0.8	5.79	0.01	0.78	6.14	0.00
20.5	3	1.0	4.84	0.70	0.98	5.15	0.90
20.6	2	1.2	3.88	0.91	1.19	4.10	1.08
20.7	2	1.4	2.99	0.33	1.40	3.13	0.40
20.8	2	1.6	2.22	0.02	1.61	2.27	0.03
20.9	2	1.8	1.58	0.11	1.82	1.58	0.11
21.0	4	2.0	1.08	7.90	2.03	1.06	8.15
21.1	0	2.2	0.71	0.71	2.24	0.68	0.68
21.2	1	2.4	0.45	0.67	2.44	0.43	0.76
SUM	100		98.31	20.90		98.67	21.33

$\chi^2_{\nu_0} = 20.90/24 = 0.87$

$\chi^2_{\nu} = 21.33/22 = 0.97$

can only expect a probability of about 0.5 for any random set of data, we conclude that this is a reasonable fit.

If we compare the data with the estimated distribution ϕ determined from the estimated mean $\bar{x} = 20.028$ cm and standard deviation $s = 0.48$ cm, we get the similar calculation in the last three columns of Table 5-2. The normalization is different from the parent distribution because the estimated standard deviation s is different from σ. With $z = x/s = 0.1/0.48 = 0.208$, the normalization for ϕ is

$$\phi = 20.8 P_G(z,0,1) \qquad z = \frac{|x - \bar{x}|}{s}$$

The number of degrees of freedom is $\nu = n - 2 = 22$ since the mean and standard deviation were both determined from the data. The resulting value of the reduced chi-square is $\chi_\nu^2 = 0.97$, which corresponds to a probability of 0.5 for 22 degrees of freedom. If we did not have previous knowledge of the parent distribution, we would be satisfied with the fit of the estimated distribution.

Note that we have included in the calculation values for x_j of 19.0 cm and 21.1 cm even though no such values were observed. In general, the fit is made only to those values which are actually observed, but for a reasonable physical case, the inclusion of such values within the observed range does not affect the results appreciably.

SUMMARY

Weighted mean:

$$\bar{x} = \frac{\Sigma(x_i/\sigma_i^2)}{\Sigma(1/\sigma_i^2)} \xrightarrow[\sigma_i = \sigma]{} \frac{1}{N}\Sigma x_i$$

Variance of mean:

$$\sigma_\mu^2 = \frac{1}{\Sigma(1/\sigma_i^2)} \xrightarrow[\sigma_i = \sigma]{} \frac{\sigma^2}{N}$$

Instrumental uncertainties: Fluctuations in measurements due to finite precision of instruments used in making measurements.

$$\sigma^2 \simeq s^2 = \frac{1}{N-1} \Sigma(x_i - \bar{x})^2$$

Statistical fluctuations: Fluctuations in observations resulting from statistical probability of taking random samples of finite numbers of items.

$$\sigma^2 = \mu \simeq \bar{x}$$

χ^2 *test:* Comparison of observed frequency distribution $f(x_j)$ of possible observations x_j vs. predicted distribution $NP(x_j)$ where N is the number of data points and $P(x_j)$ is the theoretical probability distribution.

$$\chi^2 = \sum_{j=1}^{n} \frac{[f(x_j) - NP(x_j)]^2}{NP(x_j)}$$

Degrees of freedom ν: Number of data points N minus number of parameters of distribution determined from those data points.

Reduced χ^2: $\chi_\nu^2 = \chi^2/\nu$. For χ^2 tests, χ_ν^2 should be approximately equal to 1.

Graphs and tables of χ^2: Table C-4 gives the probability that a random sample of data when compared with its *parent distribution* would yield a value of χ_ν^2 as large as the observed value.

EXERCISES

5-1 Evaluate the standard deviation and probable error of the mean in Example 2-1. Compare with the actual error. How probable is such a discrepancy?

5-2 From the data of Exercise 2-1, calculate the standard deviation σ and the uncertainty in the mean σ_μ. Is the actual error reasonable?

5-3 Do Exercise 5-2 for the data of Exercise 2-2.

5-4 For the data of Example 2-1, calculate $\chi_{\nu_0}^2$ and χ_ν^2 omitting the values for x_j of 19.0 cm and 21.1 cm from the calculation (see Table 5-3).

5-5 Find the mean, standard deviation, and standard deviation of the mean for the following data. (The third column contains data with smaller uncertainties.)

Trial	$x(\sigma)$	Trial	$x(\sigma)$	Trial	$x(\tfrac{1}{2}\sigma)$
1	1.65	11	2.69	21	2.44
2	2.00	12	1.79	22	1.75
3	1.94	13	1.62	23	1.69
4	2.13	14	1.45	24	2.06
5	1.79	15	2.05	25	1.79
6	2.11	16	1.50	26	2.01
7	1.92	17	1.21	27	1.79
8	2.48	18	2.55	28	1.69
9	2.50	19	2.05	29	2.26
10	2.05	20	1.82	30	1.97

5-6 What is the significance of Figure 5-1? Try looking at it from the lower right-hand corner as if peering up into a tunnel.

5-7 The number of gamma rays from a source and the number of background cosmic rays detected per minute are tabulated below. What is the counting rate from the source alone? What is the uncertainty in that value?

Trial	Source	Trial	Background
1	161	1	23
2	200	2	12
3	179	3	15
4	207	4	14
5	163	5	14

5-8 Do a χ^2 test between a binomial distribution with $n = 10$, $p = \tfrac{1}{2}$, and a Gaussian distribution with $\mu = 5$, $\sigma = 1.58$. Compare Figures 3-1 and 3-5.

5-9 What would be a reasonable value of χ^2 for a fit to data with a Poisson curve and $n = 10$ data points? What is the number of degrees of freedom ν for such a fit?

5-10 Roll two dice and determine the relationship between the frequency f of occurrence of 7 for the sum and the expected frequency $\langle f \rangle = N/6$. Show that $(f - \langle f \rangle)^2 \simeq \sigma_f^2 = \langle f \rangle = N/6$. Use this relationship to justify the argument that the sum in Equation (5-19) should yield a value of $\chi^2 \simeq n$.

CHAPTER SIX

LEAST-SQUARES FIT
TO A STRAIGHT LINE

6-1 DEPENDENT AND INDEPENDENT VARIABLES

It often happens that we wish to determine one characteristic of an experiment y as a function of some other quantity x. That is, we wish to find the function f such that $y = f(x)$. Instead of making a number of measurements of the quantity y for one particular value of x, we make a series of N measurements y_i, one for each of several values of the quantity $x = x_i$ where i is an index that runs from 1 to N to identify the measurements. We indicate which pair of values (x_i, y_i) corresponds to each other by using the same subscripts; for example, the nth measurement y_n is made for a value of x equal to $x = x_n$.

Table 6-1 Experimental data for a determination of the temperature T along a metal rod suspended between two constant temperature baths as a function of the position x along the rod

Trial i	Position x_i, cm	Temperature T_i, °C	x_i^2	$x_i T_i$	T_i^2	$a + bx_i$	Δ_i^2
1	1.0	15.6	1.0	15.6	243.36	14.2	1.96
2	2.0	17.5	4.0	35.0	306.25	23.6	37.21
3	3.0	36.6	9.0	109.8	1339.56	33.0	12.96
4	4.0	43.8	16.0	175.2	1918.44	42.4	1.96
5	5.0	58.2	25.0	291.0	3387.24	51.8	40.96
6	6.0	61.6	36.0	369.6	3794.56	61.2	0.16
7	7.0	64.2	49.0	449.4	4121.64	70.6	40.96
8	8.0	70.4	64.0	563.2	4956.16	80.0	92.16
9	9.0	98.8	81.0	889.2	9761.44	89.4	88.36
SUM	45.0	466.7	285.0	2898.0	29,828.65		316.69

$$\Delta = N\Sigma x_i^2 - (\Sigma x_i)^2 = 9(285) - (45)^2 = 540$$

$$a = \frac{\Sigma x_i^2 \Sigma T_i - \Sigma x_i \Sigma x_i T_i}{\Delta} = \frac{285(466.7) - 45(2898)}{540} = 4.8$$

$$b = \frac{N\Sigma x_i T_i - \Sigma x_i \Sigma T_i}{\Delta} = \frac{9(2898) - 45(466.7)}{540} = 9.4$$

$$s^2 = \frac{1}{N-2}\Sigma(T_i - a - bx_i)^2 = \frac{1}{N-2}\Sigma(\Delta_i)^2 = \tfrac{1}{7}(316.69) = 45.24$$

$$s^2 = \frac{1}{N-2}[\Sigma T_i^2 + Na^2 + b^2\Sigma x_i^2 - 2(a\Sigma T_i - ab\Sigma x_i + b\Sigma x_i T_i)]$$
$$= \tfrac{1}{7}[29,828.65 + 207.36 + 25,182.60 - 2(2240.16 - 2030.40 + 27,241.2)] = 45.24$$

$$\sigma_a^2 \simeq \frac{s^2\Sigma x_i^2}{\Delta} = 23.9 \qquad \sigma_a \simeq 4.9°C$$

$$\sigma_b^2 \simeq \frac{Ns^2}{\Delta} = 0.754 \qquad \sigma_b \simeq 0.87°C/cm$$

$$\sigma \simeq \sqrt{s^2} = 6.7 \qquad \text{P.E.} = \tfrac{2}{3}\sigma \simeq 4.5°C$$

EXAMPLE 6-1   We wish to determine the variation of the temperature T along a metal rod which is suspended between two constant temperature baths, one at a temperature $T = 0°C$ and the other at $T = 100°C$. We are interested only in the variation

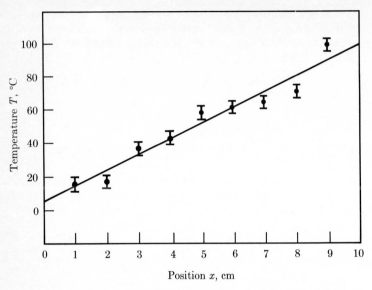

FIGURE 6-1 Graph of temperature T vs. position x for the data of Table 6-1. Probable error for data points is P.E. = 4.5°C ($\sigma \simeq s =$ 6.7°C). Straight-line fit is for $T = 4.8°C + 9.4x$ °C/cm.

of the temperature T along the rod as a function of linear position x. Table 6-1 gives the results of a series of measurements of the temperature T_i made at several positions x_i along the rod where the position x is measured from the bath at temperature $T = 0°C$. For this set of data, the subscript i runs from 1 to 9; that is, measurements of the temperature were made at each of nine positions x_i.

Figure 6-1 shows these data plotted in such a way as to show the relationship between temperature and position most clearly. The position x is considered to be the independent variable, and the temperature T is considered to be a dependent function of position. Thus, the temperatures T_i are plotted along the y axis, and the positions x_i are plotted along the x axis. From these measurements, we can determine the linear function $T(x)$, shown as a solid line, which describes the way in which the temperature T

varies as a function of position x along the rod, assuming the variation of T is linear with x as we would expect at thermal equilibrium.

EXAMPLE 6-2   Consider a counting experiment in which we count the number of events recorded in a detector as a function of time. We have a source which is emitting radiation, and the number of counts per unit time from our detector is a measure of the rate at which this radiation is being emitted. We observe qualitatively that the rate of emission is decreasing approximately linearly with time and we wish to describe this quantitatively.

We cannot determine the counting rate instantaneously because no counts will be detected in an infinitesimal time interval. But we can determine the number of counts C detected over a time interval Δt, and this should be representative of the average counting rate over that interval. Table 6-2 gives the results of a series of measurements of the number of counts C_i recorded during a fixed interval of time Δt_i for a number of intervals starting at times $t = t_i$. It is customary and convenient to make the intervals equally spaced in time as well as equally long.

In this example, the intervals are both equal $\Delta t_i = \Delta t$ and contiguous $\Delta t_i = t_{i+1} - t_i$; the times t_i at which the successive intervals start are given by $t_i = (i - 1)\, \Delta t$, with time measured from the beginning of the first interval.

Figure 6-2 shows these data plotted as a multichannel spectrum, with the number of counts per interval C_i plotted as a function of the starting time t_i for each interval Δt. The solid line represents the function $C = f(t)$ which is a linear approximation to the rate at which the source of counts is decreasing.

Linear approximation   In both of these examples, the functional relationship between the variables y and x (that is, between the quantities T and x or C and t) can be approximated by a straight line. We will consider in this chapter a method for determining the most probable values for the coefficients a and b, considering the functional relationship $y = f(x)$ to be linear and

Table 6-2 Experimental data for a determination of the counting rate C in a detector as a function of time t (the starting times for the intervals Δt during which the counts are accumulated)

Trial i	Time t_i, sec	Counts C_i, per 15 sec	$\dfrac{1}{C_i}$	$\dfrac{t_i}{C_i}$	$\dfrac{t_i^2}{C_i}$	$a + bt_i$
1	0	106	0.00944	0.0	0.0	104.4
2	15	80	0.01250	0.188	2.81	95.2
3	30	98	0.01020	0.306	9.18	86.1
4	45	75	0.01333	0.600	27.00	76.9
5	60	74	0.01351	0.811	48.65	67.8
6	75	73	0.01370	1.027	77.06	58.6
7	90	49	0.02041	1.837	165.31	49.5
8	105	38	0.02632	2.763	290.13	40.3
9	120	37	0.02703	3.243	389.19	31.2
10	135	22	0.04546	6.136	828.41	22.0
SUM	675	652	0.19190	16.911	1837.74	

$$\Delta = \sum \frac{t_i^2}{C_i} \sum \frac{1}{C_i} - \left(\sum \frac{t_i}{C_i} \right)^2 = 1837.74(.1919) - (16.911)^2 = 66.66$$

$$a = \frac{N\Sigma(t_i^2/C_i) - \Sigma t_i \Sigma(t_i/C_i)}{\Delta} = \frac{10(1837.74) - 675(16.911)}{66.66} = 104.4$$

$$b = \frac{\Sigma t_i \Sigma(1/C_i) - N\Sigma(t_i/C_i)}{\Delta} = \frac{675(.1919) - 10(16.911)}{66.66} = -.61/\text{sec}$$

$$\sigma_a^2 \simeq \frac{\Sigma(t_i^2/C_i)}{\Delta} = \frac{1837.74}{66.66} = 27.6 \qquad \sigma_a \simeq 5.3$$

$$\sigma_b^2 \simeq \frac{\Sigma(1/C_i)}{\Delta} = \frac{.1919}{66.66} = 0.00288 \qquad \sigma_b \simeq 0.054/\text{sec}$$

of the form

$$y = a + bx$$

We cannot fit a straight line to the data exactly in either example because it is impossible to draw a straight line through all the points. For a set of N arbitrary points, it is theoretically possible to fit a polynomial of degree $N - 1$ exactly, but for our experiments, the coefficients of higher-order terms would have questionable physical significance. The fluctuations of the individual points above and below the solid curves are assumed to be

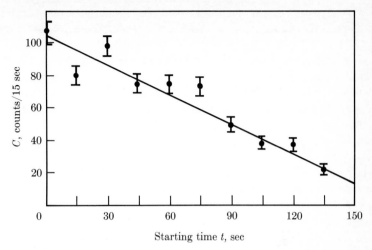

FIGURE 6-2 Graph of the number of counts C per 15 seconds in a detector vs. the starting time t of each 15-sec interval for the data of Table 6-2. Probable error for data points from statistics P.E. = $0.67\sigma_i$ ($\sigma_i^2 = 1/C_i$). Straight-line fit is for $C = 104.4 - 0.61t$/sec.

due to experimental uncertainties in the individual measurements and not to any physical relationship. Such an assumption should be justified on theoretical grounds and verified by investigating the reproducibility of individual measurements. In Chapter 7 we will develop a method for testing whether higher-order terms are significant.

If we were to make a series of measurements of the dependent quantity y for one particular value of the independent quantity x, for example, we would find that the measured values were distributed about a mean in the same manner as discussed in Chapter 3. Any one measurement y_i is expected to differ from that mean by about 1 standard deviation σ_i more or less. By making a number of measurements for each value of the independent quantity x_i, we could determine mean values $\overline{y_i}$ with any desired precision. We will, however, make only one measurement y_i for each value of $x = x_i$, and, therefore, we will determine the value of y corresponding to that value of x with an uncertainty, which is characterized by the standard deviation σ_i of the distribution for data at that point.

Uncertainties We will assume for simplicity in all the following discussions that we can ascribe all the uncertainty of measurement to the dependent parameter. This is equivalent to assuming that the precision of the determination of x is considerably higher than that of y. Thus, we will consider our independent quantity to be determined exactly and our dependent quantity to be determined with some uncertainty. This difference is illustrated in Figures 6-1 and 6-2 by the fact that the uncertainties are indicated by vertical lines for the dependent variables and not for the independent variables.

If this assumption is to be true for Example 6-1, then we must determine the position along the rod with higher precision than that of our determination of the temperature. Quantitatively, we express the relative uncertainty of each quantity in proportion to the range through which that quantity changes and compare these two terms

$$\frac{\sigma_x}{x_i - x_j} \ll \frac{\sigma_T}{T_i - T_j}$$

where σ_x and σ_T are standard deviations for measurements of x and T, respectively, and the subscripts i and j correspond to two representative measurements.

Similarly, for Example 6-2, we must be able to determine the time for each interval with a higher precision than that for determining the number of counts in each interval. To put it quantitatively

$$\frac{\sigma_t}{\Delta t} \ll \frac{\sigma_i}{C_i - C_{i-1}}$$

where σ_t and σ_i are the standard deviations for determinations of the time intervals Δt and the number of counts in the intervals C_i, respectively.

The left-hand uncertainty refers only to the uncertainty in determining the starting times t_i, however. There are also uncertainties $\sigma_{\Delta t}$ in the durations of the intervals which result in experimental uncertainties σ_{ei} in the observed counts C_i.

$$\sigma_{ei} = \sigma_{\Delta t} \frac{C_i}{\Delta t}$$

These must be combined with the statistical uncertainties σ_{si} expected from statistical fluctuations (see Section 6-4) to yield the total uncertainty σ_i. According to Equation (4-9), the total uncertainty is the root sum square of the contributing uncertainties.

$$\sigma_i^2 = \sigma_{ei}^2 + \sigma_{si}^2 \qquad (6\text{-}1)$$

We are not always justified in ascribing all the uncertainties to the dependent parameter. Sometimes the uncertainties in the determinations of both quantities x and y are nearly equal. But our fitting procedure will still be fairly accurate if we combine the uncertainties of both quantities and assign them to the dependent variable alone. This is the assumption we will generally make. We will proceed with our calculations on the basis that we need consider uncertainties in the dependent quantity only, but we will understand that the corresponding fluctuations were originally derived from uncertainties in the determinations of both quantities.

In those cases where the uncertainties in the determination of the independent quantity are considerably greater than those of the dependent quantity, we might benefit from redefining which quantity we consider as dependent and which as independent.

6-2 METHOD OF LEAST SQUARES

Our data consist of pairs of measurements (x_i, y_i) of an independent variable x and a dependent variable y. We wish to fit the data with an equation of the form

$$y = a + bx \qquad (6\text{-}2)$$

by determining the values of the coefficients a and b such that the discrepancy is minimized between the values of our measurements y_i and the corresponding values $y = f(x_i)$ given by Equation (6-2). We cannot determine the coefficients exactly with only a finite number of observations, but we do want to extract from these data the most probable estimates for the coefficients.

The problem is to establish criteria for *minimizing the discrepancy* and optimizing the estimates of the coefficients. For any arbitrary values of a and b, we can calculate the deviations Δy_i between each of the observed values y_i and the corresponding calculated values

$$\Delta y_i = y_i - a - b x_i$$

If the coefficients are well chosen, these deviations should be relatively small. The sum of these deviations is not a good measure of how well we have approximated the data with our calculated straight line because large positive deviations can be balanced by large negative ones to yield a small sum even when the fit is bad. We might consider instead summing the absolute values of the deviations, but this leads to difficulties in obtaining an analytical solution. We might consider instead the sum of the squares of deviations.

There is no unique correct method for optimizing the coefficients which is valid for all cases. There exists, however, a method which can be fairly well justified, which is simple and straightforward, which is well established experimentally as being appropriate, and which is accepted by convention. This is the method of least squares, similar to that discussed in Section 5-1 but extrapolated to include more than one variable.

Method of maximum likelihood Our data consist of a sample of observations extracted from a parent distribution which determines the probability of making any particular observation. Let us define parent coefficients a_0 and b_0 such that the actual relationship between y and x is given by

$$y(x) = a_0 + b_0 x \qquad (6\text{-}3)$$

For any given value of $x = x_i$, we can calculate the probability P_i for making the observed measurement y_i, assuming a Gaussian distribution with a standard deviation σ_i for the observations about the actual value $y(x_i)$.

$$P_i = \frac{1}{\sigma_i \sqrt{2\pi}} \exp\left\{ -\frac{1}{2} \left[\frac{y_i - y(x_i)}{\sigma_i} \right]^2 \right\}$$

The probability for making the observed set of measurements of the N values of y_i is the product of these probabilities

$$P(a_0, b_0) = \Pi P_i = \prod \left(\frac{1}{\sigma_i \sqrt{2\pi}} \right) \exp\left\{ -\frac{1}{2} \sum \left[\frac{y_i - y(x_i)}{\sigma_i} \right]^2 \right\} \tag{6-4}$$

where the product Π is taken for i ranging from 1 to N.

Similarly, for any estimated values of the coefficients a and b, we can calculate the probability that we should make the observed set of measurements

$$P(a,b) = \prod \left(\frac{1}{\sigma_i \sqrt{2\pi}} \right) \exp\left[-\frac{1}{2} \sum \left(\frac{\Delta y_i}{\sigma_i} \right)^2 \right] \tag{6-5}$$

The *method of maximum likelihood* consists of making the assumption that the observed set of measurements is more likely to have come from the actual parent distribution of Equation (6-3) than from any other similar distribution with different coefficients and, therefore, that the probability of Equation (6-4) is the maximum probability attainable with Equation (6-5). The best estimates for a and b are therefore those values which maximize the probability of Equation (6-5).

The first term of Equation (6-5) is a constant, independent of the values of a or b. Thus, maximizing the probability $P(a,b)$ is equivalent to minimizing the sum in the exponential. We define the quantity χ^2 to be this sum

$$\chi^2 \equiv \sum \left(\frac{\Delta y_i}{\sigma_i} \right)^2 = \sum \left[\frac{1}{\sigma_i^2} (y_i - a - bx_i)^2 \right] \tag{6-6}$$

and consider this to be the appropriate measure of the goodness of fit. We have used the same symbol χ^2 defined earlier in Equation (5-19) because this is essentially the same definition in a different context.

Our method for finding the optimum fit to the data will be to minimize this weighted sum of squares of deviations χ^2 and, hence, to find the fit which produces the smallest sum of squares or the *least-squares fit*.

6-3 INSTRUMENTAL UNCERTAINTIES

If the quantity y is one which can be measured with a physical instrument, the uncertainty in each measurement generally comes from fluctuations in repeated readings of the instrumental scale, either because the settings are not exactly reproducible due to imperfections in the equipment, or because of human imprecision in observing the settings, or a combination of both. Such uncertainties are called *instrumental* because they arise from a lack of perfect precision in the measuring instruments (including the observer).

We can include in this category experiments which deal with measurements of such characteristics as length, mass, voltage, current, etc. In the discussion which follows, we shall consider first the simpler case where the absolute uncertainties are equal throughout the entire experiment. Later we shall consider the refinement of utilizing the standard deviation as a weighting factor corresponding to the precision which may vary from one part of the experiment to another as when the scale factor is changed or the scale is non-linear. In the next section we will consider uncertainties resulting from statistical fluctuations rather than from experimental precision.

For example, we include in this category such experiments as that of Example 6-1 illustrated in Figure 6-1 in which the observed quantity is the temperature T, measured with a thermometer consisting of a thermocouple and a meter with a linear scale. The fluctuations in the data result from errors in reading the meter, and these errors are just as large for readings near the low end of the scale as for readings near the high end (ignoring errors in calibration). So long as we do not change the scale to measure temperatures outside the reasonable range, the absolute values (rather than the relative values) of the uncertainties will be the same for all measurements.

Minimizing χ^2 In order to find the values of the coefficients a and b which yield the minimum value for χ^2, we use the method of calculus described in Appendix A in the same way as in Section 5-1, extrapolated to minimizing the function with respect to more than one coefficient. The minimum value of the function χ^2 of Equation (6-6) is one which yields a value of 0 for both of the partial derivatives with respect to each of the coefficients

$$
\begin{aligned}
\frac{\partial}{\partial a} \chi^2 &= \frac{\partial}{\partial a} \left[\frac{1}{\sigma^2} \Sigma(y_i - a - bx_i)^2 \right] \\
&= \frac{-2}{\sigma^2} \Sigma(y_i - a - bx_i) = 0 \\
\frac{\partial}{\partial b} \chi^2 &= \frac{\partial}{\partial b} \left[\frac{1}{\sigma^2} \Sigma(y_i - a - bx_i)^2 \right] \\
&= \frac{-2}{\sigma^2} \Sigma[x_i(y_i - a - bx_i)] = 0
\end{aligned}
\tag{6-7}
$$

where we have for the present considered all of the standard deviations equal $\sigma_i = \sigma$.

These equations can be rearranged to yield a pair of simultaneous equations

$$
\begin{aligned}
\Sigma y_i &= \Sigma a + \Sigma b x_i = aN + b\Sigma x_i \\
\Sigma x_i y_i &= \Sigma a x_i + \Sigma b x_i^2 = a\Sigma x_i + b\Sigma x_i^2
\end{aligned}
\tag{6-8}
$$

where we have substituted N for $\Sigma(1)$ since the sum runs for $i = 1$ to N. This development is discussed more fully in Appendix A.

We wish to solve Equations (6-8) for the coefficients a and b. This will give us the values of the coefficients for which χ^2, the

Program 6-1   LINFIT   Least-squares fit to a straight line.

```
C      SUBROUTINE LINFIT
C
C      PURPOSE
C        MAKE A LEAST-SQUARES FIT TO DATA WITH A STRAIGHT LINE
C          Y = A + B*X
C
C      USAGE
C        CALL LINFIT (X, Y, SIGMAY, NPTS, MODE, A, SIGMAA, B, SIGMAB, R)
C
C      DESCRIPTION OF PARAMETERS
C        X      - ARRAY OF DATA POINTS FOR INDEPENDENT VARIABLE
C        Y      - ARRAY OF DATA POINTS FOR DEPENDENT VARIABLE
C        SIGMAY - ARRAY OF STANDARD DEVIATIONS FOR Y DATA POINTS
C        NPTS   - NUMBER OF PAIRS OF DATA POINTS
C        MODE   - DETERMINES METHOD OF WEIGHTING LEAST-SQUARES FIT
C                 +1 (INSTRUMENTAL) WEIGHT(I) = 1./SIGMAY(I)**2
C                  0 (NO WEIGHTING) WEIGHT(I) = 1.
C                 -1 (STATISTICAL)  WEIGHT(I) = 1./Y(I)
C        A      - Y INTERCEPT OF FITTED STRAIGHT LINE
C        SIGMAA - STANDARD DEVIATION OF A
C        B      - SLOPE OF FITTED STRAIGHT LINE
C        SIGMAB - STANDARD DEVIATION OF B
C        R      - LINEAR CORRELATION COEFFICIENT
C
C      SUBROUTINES AND FUNCTION SUBPROGRAMS REQUIRED
C        NONE
C
C      MODIFICATIONS FOR FORTRAN II
C        OMIT DOUBLE PRECISION SPECIFICATIONS
C        CHANGE DSQRT TO SQRTF IN STATEMENTS 67, 68, AND 71
C
```

sum of squares of the deviations of the data points from the calculated fit, is a minimum. The solution can be found in any one of a number of different ways, but, for generality for later similar but more complex situations, let us use the method of determinants. Appendix B contains a discussion of this method and gives the rules for obtaining a solution for any number of simultaneous equations.

The solutions are:

$$a = \frac{1}{\Delta} \begin{vmatrix} \Sigma y_i & \Sigma x_i \\ \Sigma x_i y_i & \Sigma x_i{}^2 \end{vmatrix} = \frac{1}{\Delta} (\Sigma x_i{}^2 \Sigma y_i - \Sigma x_i \Sigma x_i y_i)$$

$$b = \frac{1}{\Delta} \begin{vmatrix} N & \Sigma y_i \\ \Sigma x_i & \Sigma x_i y_i \end{vmatrix} = \frac{1}{\Delta} (N \Sigma x_i y_i - \Sigma x_i \Sigma y_i) \qquad (6\text{-}9)$$

$$\Delta = \begin{vmatrix} N & \Sigma x_i \\ \Sigma x_i & \Sigma x_i{}^2 \end{vmatrix} = N \Sigma x_i{}^2 - (\Sigma x_i)^2$$

Table 6-1 shows a sample calculation for the data of our first

Program 6-1 LINFIT (*continued*)

```
      SUBROUTINE LINFIT (X,Y,SIGMAY,NPTS,MODE,A,SIGMAA,B,SIGMAB,R)
      DOUBLE PRECISION SUM, SUMX, SUMY, SUMX2, SUMXY, SUMY2
      DOUBLE PRECISION XI, YI, WEIGHT, DELTA, VARNCE
      DIMENSION X(1), Y(1), SIGMAY(1)
C
C        ACCUMULATE WEIGHTED SUMS
C
   11 SUM = 0.
      SUMX = 0.
      SUMY = 0.
      SUMX2 = 0.
      SUMXY = 0.
      SUMY2 = 0.
   21 DO 50 I=1, NPTS
      XI = X(I)
      YI = Y(I)
      IF (MODE) 31, 36, 38
   31 IF (YI) 34, 36, 32
   32 WEIGHT = 1. / YI
      GO TO 41
   34 WEIGHT = 1. / (-YI)
      GO TO 41
   36 WEIGHT = 1.
      GO TO 41
   38 WEIGHT = 1. / SIGMAY(I)**2
   41 SUM   = SUM   + WEIGHT
      SUMX  = SUMX  + WEIGHT*XI
      SUMY  = SUMY  + WEIGHT*YI
      SUMX2 = SUMX2 + WEIGHT*XI*XI
      SUMXY = SUMXY + WEIGHT*XI*YI
      SUMY2 = SUMY2 + WEIGHT*YI*YI
   50 CONTINUE
C
C        CALCULATE COEFFICIENTS AND STANDARD DEVIATIONS
C
   51 DELTA = SUM*SUMX2 - SUMX*SUMX
      A = (SUMX2*SUMY - SUMX*SUMXY) / DELTA
   53 B = (SUMXY*SUM  - SUMX*SUMY ) / DELTA
   61 IF (MODE) 62, 64, 62
   62 VARNCE = 1.
      GO TO 67
   64 C = NPTS - 2
      VARNCE = (SUMY2 + A*A*SUM + B*B*SUMX2
     1 -2.*(A*SUMY + B*SUMXY - A*B*SUMX)) / C
   67 SIGMAA = DSQRT(VARNCE*SUMX2 / DELTA)
   68 SIGMAB = DSQRT(VARNCE*SUM  / DELTA)
   71 R = (SUM*SUMXY - SUMX*SUMY) /
     1 DSQRT(DELTA*(SUM*SUMY2 - SUMY*SUMY))
      RETURN
      END
```

example. The calculation is straightforward, though tedious. We accumulate four sums (Σx_i, $\Sigma y_i = \Sigma T_i$, Σx_i^2, and $\Sigma x_i y_i = \Sigma x_i T_i$), and combine them according to Equations (6-9) to find numerical values for a and b.

Program 6-1 The same method of calculation is also illustrated with the computer routine LINFIT of Program 6-1.

This is a Fortran subroutine to calculate the coefficients a and b for a least-squares fit of a straight line to an array of data points for any one of three different experimental conditions. The input variables are X, Y, SIGMAY, NPTS, and MODE, and the output variables are A, SIGMAA, B, SIGMAB, and R.

For the calculation discussed above, following the method of Equations (6-9), the variable MODE must have the value 0. This indicates to the subroutine that we have not considered any weighting of the fitting procedure by including the standard deviations of individual points. The variable NPTS represents the number of pairs of data points NPTS = N. The independent quantities x_i are assumed to be stored in the array X, and the dependent data points are assumed to be stored in the array Y, with the ordering identical for the two arrays. The array SIGMAY may be ignored; in this mode the subroutine does not use it or modify it.

The four sums given above (Σx_i, Σy_i, Σx_i^2, and $\Sigma x_i y_i$) are accumulated in statements 41–50 as part of the DO loop starting at statement 21. The variable WEIGHT is given a value of 1 in statement 36 and can be ignored. The calculations of Equations (6-9) are carried out in statements 51–53, with N replaced by SUM = $\Sigma(1)$. The coefficients A = a and B = b are returned to the calling program as arguments of the calling sequence. The remainder of the subroutine pertains to material not yet discussed.

Weighting the fit   If the fluctuations in the data are due to instrumental errors, but for reasons of scale changes, non-linear scales, etc., the uncertainties are not equal throughout, it is necessary to reintroduce the standard deviation from Equation (6-6) as a weighting factor into Equations (6-7) to (6-9). Instead of minimizing the simple sum of the squares of deviations as in Equations (6-7), we weight each term of the sum in χ^2 according to how large or small the deviation is expected to be at that point before summing.

Minimizing χ^2 as given in Equation (6-6), Equation (6-7) becomes

$$\frac{\partial}{\partial a}\chi^2 = \frac{\partial}{\partial a}\sum\left[\frac{1}{\sigma_i{}^2}(y_i - a - bx_i)^2\right]$$

$$= -2\sum\left[\frac{1}{\sigma_i{}^2}(y_i - a - bx_i)\right] = 0$$

$$\frac{\partial}{\partial b}\chi^2 = \frac{\partial}{\partial b}\sum\left[\frac{1}{\sigma_i{}^2}(y_i - a - bx_i)^2\right]$$

$$= -2\sum\left[\frac{x_i}{\sigma_i{}^2}(y_i - a - bx_i)\right] = 0$$

(6-10)

These equations can be rearranged to yield a pair of simultaneous equations analogous to Equations (6-8).

$$\sum\frac{y_i}{\sigma_i{}^2} = a\sum\frac{1}{\sigma_i{}^2} + b\sum\frac{x_i}{\sigma_i{}^2}$$

$$\sum\frac{x_i y_i}{\sigma_i{}^2} = a\sum\frac{x_i}{\sigma_i{}^2} + b\sum\frac{x_i{}^2}{\sigma_i{}^2}$$

(6-11)

The solutions are similar to Equations (6-9).

$$a = \frac{1}{\Delta}\begin{vmatrix} \sum\dfrac{y_i}{\sigma_i{}^2} & \sum\dfrac{x_i}{\sigma_i{}^2} \\ \sum\dfrac{x_i y_i}{\sigma_i{}^2} & \sum\dfrac{x_i{}^2}{\sigma_i{}^2} \end{vmatrix} = \frac{1}{\Delta}\left(\sum\frac{x_i{}^2}{\sigma_i{}^2}\sum\frac{y_i}{\sigma_i{}^2} - \sum\frac{x_i}{\sigma_i{}^2}\sum\frac{x_i y_i}{\sigma_i{}^2}\right)$$

$$b = \frac{1}{\Delta}\begin{vmatrix} \sum\dfrac{1}{\sigma_i{}^2} & \sum\dfrac{y_i}{\sigma_i{}^2} \\ \sum\dfrac{x_i}{\sigma_i{}^2} & \sum\dfrac{x_i y_i}{\sigma_i{}^2} \end{vmatrix}$$

$$= \frac{1}{\Delta}\left(\sum\frac{1}{\sigma_i{}^2}\sum\frac{x_i y_i}{\sigma_i{}^2} - \sum\frac{x_i}{\sigma_i{}^2}\sum\frac{y_i}{\sigma_i{}^2}\right)$$

(6-12)

$$\Delta = \begin{vmatrix} \sum\dfrac{1}{\sigma_i{}^2} & \sum\dfrac{x_i}{\sigma_i{}^2} \\ \sum\dfrac{x_i}{\sigma_i{}^2} & \sum\dfrac{x_i{}^2}{\sigma_i{}^2} \end{vmatrix} = \sum\frac{1}{\sigma_i{}^2}\sum\frac{x_i{}^2}{\sigma_i{}^2} - \left(\sum\frac{x_i}{\sigma_i{}^2}\right)^2$$

Such a calculation is even more tedious than that of Equations (6-9) and presupposes a knowledge of the magnitudes of the standard deviations σ_i for each of the data points. Fortunately, we are generally only interested in the relative uncertainties of various sections of our data when we modify scale factors, etc. For such purposes, the standard deviations can have any arbitrary overall normalization (e.g., the smallest value of $\sigma_i{}^2$ may be set equal to 1 so that all the other values are integers).

The method of calculation for the general case when the values of σ_i are known is illustrated in the subroutine LINFIT with the variable MODE given a value of $+1$ (or any positive integer). The standard deviations σ_i must be stored in the array SIGMAY with the same ordering as the data in arrays X and Y. The calculation is the same as for the earlier example, except that the variable WEIGHT is given the value 1. / $(SIGMAY(I))^2$ in statement 38 for each term so that the sums accumulated are those required for Equations (6-12).

6-4 STATISTICAL FLUCTUATIONS

If the quantity y represents the number of counts in a detector per unit time interval, as in Example 6-2, then it is generally true that the uncertainty in each measurement y_i is directly related to the magnitude of y (as discussed in Section 3-2), and, therefore, the standard deviations σ_i associated with these measurements cannot be considered equal over any reasonable range of values. Such uncertainties are called *statistical* because they arise not from a lack of perfect precision in the measuring instruments, but from statistical fluctuations in the collections of finite numbers of counts over finitely long intervals of time.

In our counting experiment of Example 6-2, for example, we would expect from the straight-line fit to the data that we should receive about 100 counts in our detector during the first time interval. What we mean by this is that if the counting rate were continued indefinitely at the same rate for a large number of intervals, the average number of counts received per interval would be very nearly 100. Since the counts are distributed randomly in time, however, we would expect to receive more than 100 counts in some intervals and fewer than 100 in others. The fluctuations of the number of counts actually received in each interval around the average number of counts are statistical fluctuations related to the probability of receiving more or fewer than the average number of counts in any time interval.

There can be instrumental uncertainties as well contributing to the overall uncertainties. We can determine the time intervals with only finite precision, and the same precision applies to the determination of the starting times t_i, though this is generally a negligible correction. These are actually uncertainties in the independent quantity x, but we have agreed to assign them arbitrarily to the dependent quantity y. For counting experiments these contributions to the overall uncertainty are generally ignored on the assumption that the statistical fluctuations dominate. Where this is not true, the standard deviations to be used in Equations (6-10) to (6-12) as weighting factors must be the root sum squares of the standard deviations for the experimental deviations $\sigma_i(x_i)$ in x and the statistical deviations $\sigma_i(y_i)$ in y as given in Equation (6-1).

$$\sigma_i{}^2 = \sigma_i{}^2(x_i) + \sigma_i{}^2(y_i)$$

Estimate of σ If the fluctuations in the measurements y_i are statistical, we can estimate analytically what the standard deviation corresponding to each observation is, without having to determine it experimentally. If we were to make the same measurement repeatedly, we would find that the observed values were distributed about their mean in a Poisson distribution (as discussed in Section 3-2) instead of a Gaussian distribution. We can justify the use of this distribution intuitively by considering that we would expect a distribution which is related to the binomial distribution, but which is consistent with our boundary conditions that we may receive any positive number of counts, but no fewer than 0 counts, in any time interval.

One immediate advantage of the Poisson distribution is that the standard deviation is automatically determined.

$$\sigma = \sqrt{y} \tag{6-13}$$

The standard deviation, a measure of the absolute uncertainty of each measurement, increases as the square root of the average counting rate. The relative uncertainty, the ratio of the standard deviation to the average rate σ/y, decreases as the number of counts received per interval increases. Thus, as we expect, our relative uncertainties are smaller when our counting rates are higher.

The value for y to be used in Equation (6-13) for determining the standard deviation σ is, of course, the value of the parent population of which each measurement is only an approximate sample. In the limit of an infinite number of determinations, the average of all the measurements would very closely approximate this parent value, but we generally have no intention of making more than one measurement at each value of x, much less an infinite number. We can approximate the parent value of y by using the calculated value $y(x)$ from our fit, but that complicates the fitting procedure.

Optimizing coefficients   If we substitute the approximation $\sigma_i = y(x_i)$ into the definition of χ^2 in Equation (6-6) and minimize the value of χ^2 as in Equations (6-10), the least-squares values of a and b are given by a pair of simultaneous equations analogous to those of Equations (6-11).

$$
\begin{aligned}
N &= \sum \frac{y_i{}^2}{(a + bx_i)^2} \\
\Sigma x_i &= \sum \frac{x_i y_i{}^2}{(a + bx_i)^2}
\end{aligned}
\tag{6-14}
$$

Actually, the least-squares method is not applicable because its derivation in Section 6-2 depended on the assumption that the uncertainties were distributed according to the Gaussian distribution and that these uncertainties follow the Poisson distribution.

Let us replace the probability $P(a,b)$ of Equation (6-5) with the corresponding probability for observing y_i counts from a

Poisson distribution with means $\mu_i = y(x_i)$

$$P(a,b) = \prod_{i=1}^{N} \left\{ \frac{[y(x_i)]^{y_i}}{y_i!} e^{-y(x_i)} \right\}$$

and apply the method of maximum likelihood to this probability. It is easier and equivalent to maximize the natural logarithm of the probability

$$\ln P(a,b) = \Sigma[y_i \ln y(x_i)] - \Sigma y(x_i) + \text{const}$$

with respect to each of the coefficients a and b. The result is a pair of simultaneous equations similar to those of Equations (6-14)

$$N = \sum \frac{y_i}{a + bx_i}$$
$$\Sigma x_i = \sum \frac{x_i y_i}{a + bx_i}$$

(6-15)

but with less emphasis on fitting the larger values of y_i.

In general, however, because of the difficulty of solving Equations (6-15) and their extrapolations to higher-order terms, we will make the following two assumptions to simplify the calculation:

1. The shapes of the individual Poisson distributions governing the fluctuations in the observed y_i are nearly Gaussian.
2. Within the errors of the experiment, the uncertainties σ_i in the observations y_i may be approximated by

$$\sigma_i^2 \simeq y_i \tag{6-16}$$

It is important to note that the value of y_i to be used in Equation (6-16) is that of the raw data, i.e., the actual number of events observed, not a normalized value to conform to a given set of units. If counts are observed for several different lengths of time intervals, for example, the value of y_i to be used in accumulating sums should be normalized to a common time interval, but the value of y_i to be used in Equation (6-16) should be equal to the actual observed counts in the intervals.

With the approximation of Equation (6-16), Equation (6-12) becomes

$$a = \frac{1}{\Delta} \begin{vmatrix} N & \sum \dfrac{x_i}{y_i} \\ \Sigma x_i & \sum \dfrac{x_i{}^2}{y_i} \end{vmatrix} = \frac{1}{\Delta} \left(N \sum \frac{x_i{}^2}{y_i} - \Sigma x_i \sum \frac{x_i}{y_i} \right)$$

$$b = \frac{1}{\Delta} \begin{vmatrix} \sum \dfrac{1}{y_i} & N \\ \sum \dfrac{x_i}{y_i} & \Sigma x_i \end{vmatrix} = \frac{1}{\Delta} \left(\Sigma x_i \sum \frac{1}{y_i} - N \sum \frac{x_i}{y_i} \right) \qquad (6\text{-}17)$$

$$\Delta = \begin{vmatrix} \sum \dfrac{1}{y_i} & \sum \dfrac{x_i}{y_i} \\ \sum \dfrac{x_i}{y_i} & \sum \dfrac{x_i{}^2}{y_i} \end{vmatrix} = \sum \frac{x_i{}^2}{y_i} \sum \frac{1}{y_i} - \left(\sum \frac{x_i}{y_i} \right)^2$$

Calculations Table 6-2 shows a sample calculation for the data of Example 6-2. The calculation is similar to that in Table 6-1 but it involves accumulating weighted sums instead of simple sums. Note that there is more of a possibility in this kind of calculation of having the two terms in the expression for Δ nearly cancel, and therefore care must be taken that the final precision is high enough.

The subroutine LINFIT in Program 6-1 is also capable of performing the calculation of Equations (6-17). When the variable MODE is set equal to -1 (or any negative integer), the weighting factor is calculated in statement 32 as

$$\text{WEIGHT} = \frac{1}{\sigma_i{}^2} = \frac{1.}{\text{Y}(\text{I})}$$

It is, of course, impossible to observe a negative number of counts, and it is physically meaningless to use Poisson statistics to fit to data which have negative points.

Experimentally, however, we often wish to fit curves to data which have been obtained as the difference between observed data and estimated background. The correct method for treating such a case would be to determine a composite set of standard deviations from the root sum square of the standard deviations for the observed data and the background estimation.

$$\sigma_i{}^2 = \sigma_i{}^2(\text{DATA}) + \sigma_i{}^2(\text{BACKGROUND})$$

In practice, however, if the background is small, there is a great temptation to fit the difference data directly using Poisson statistics. For this reason, provision is made for allowing 0 or negative as well as positive values of Y(I) giving them "reasonable" weights.

6-5 ESTIMATION OF ERRORS

In order to find the uncertainty in the estimation of the coefficients a and b in our fitting procedure, we refer to our discussion of the propagation of errors in Chapter 4. Each of our data points y_i has been used in the determination of the parameters, and each has contributed some fraction of its own uncertainty to the uncertainty in our final determination. Ignoring systematic errors which would introduce correlations between the uncertainties, the standard deviation σ_z of the determination of any parameter z is given by Equation (4-9) as the root sum square of the products of the standard deviation of each data point σ_i multiplied by the effect which that data point has on the determination of the parameter z.

$$\sigma_z{}^2 = \Sigma \left[\sigma_i{}^2 \left(\frac{\partial z}{\partial y_i} \right)^2 \right] \qquad (6\text{-}18)$$

Instrumental uncertainties   If we assume that the uncertainties are instrumental and that the standard deviations σ_i for the data points y_i are all equal $\sigma_i = \sigma$, then we can estimate them from the data. Our definition from Equation (2-10) of the sample variance s^2 which approximates σ^2 is the sum of the squares of deviations of the data points from the calculated mean divided by the number of degrees of freedom. In this case, the number of degrees of freedom is the number of data points N minus the num-

ber of parameters (2) which we determined before calculating the
mean. Thus, our estimated parent standard deviation $\sigma_i = \sigma$ is

$$\sigma^2 \simeq s^2 = \frac{1}{N-2} \sum (y_i - a - bx_i)^2 \tag{6-19}$$

Note that it is this common uncertainty which we have minimized
in our least-squares fitting.

The derivatives in Equation (6-18) can be evaluated by
taking the derivatives of Equations (6-9),

$$\frac{\partial a}{\partial y_j} = \frac{1}{\Delta} (\Sigma x_i^2 - x_j \Sigma x_i)$$

$$\frac{\partial b}{\partial y_j} = \frac{1}{\Delta} (N x_j - \Sigma x_i) \tag{6-20}$$

$$\Delta = N\Sigma x_i^2 - (\Sigma x_i)^2$$

where the sums are taken over only the i indices and j is another
dummy index.

Combining Equations (6-18) and (6-20), we can find an
expression for the uncertainty in parameter a

$$\sigma_a^2 \simeq \sum_{j=1}^{N} \frac{\sigma^2}{\Delta^2} [(\Sigma x_i^2)^2 - 2x_j \Sigma x_i^2 \Sigma x_i + x_j^2 (\Sigma x_i)^2]$$

$$= \frac{\sigma^2}{\Delta^2} [N(\Sigma x_i^2)^2 - 2(\Sigma x_i)^2 \Sigma x_i^2 + \Sigma x_i^2 (\Sigma x_i)^2]$$

$$= \frac{\sigma^2}{\Delta^2} (\Sigma x_i^2)[N\Sigma x_i^2 - (\Sigma x_i)^2] = \frac{\sigma^2}{\Delta} \Sigma x_i^2 \tag{6-21}$$

where j is, again, a dummy index like i. Similarly, the uncertainty in parameter b is

$$\sigma_b^2 \simeq \sum_{j=1}^{N} \frac{\sigma^2}{\Delta^2} [N^2 x_j^2 - 2N x_j \Sigma x_i + (\Sigma x_i)^2]$$

$$= \frac{\sigma^2}{\Delta^2} [N^2 \Sigma x_i^2 - 2N(\Sigma x_i)^2 + N(\Sigma x_i)^2]$$

$$= \frac{N\sigma^2}{\Delta^2} [N\Sigma x_i^2 - (\Sigma x_i)^2] = N \frac{\sigma^2}{\Delta} \tag{6-22}$$

where σ is given by Equation (6-19).

$$\sigma^2 \simeq s^2 = \frac{1}{N-2} (\Sigma y_i^2 + Na^2 + b^2\Sigma x_i^2$$
$$- 2a\Sigma y_i - 2b\Sigma x_i y_i + 2ab\Sigma x_i) \quad (6\text{-}23)$$

Sample calculation In Table 6-1, s is calculated by two different methods: (1) by calculating the individual deviations and summing the squares as in Equation (6-19), and (2) by calculating the sum of y_i^2 and combining this sum with the sums already calculated, using the expansion of Equation (6-23) to find s. These two methods are equivalent in principle, but round-off errors in computation generally make the second method less precise. Note how many more significant figures must be retained in the terms of the sum than in the final result. The uncertainties for coefficients a and b follow the expressions in (6-21) and (6-22).

Similarly, the subroutine LINFIT calculates σ_a = SIGMAA and σ_b = SIGMAB by the second method according to the same formulas, providing MODE = 0 for instrumental uncertainties with no weights given. Statements 64–68 perform these calculations, using the sums developed in statements 41–50 for the fitting (VARNCE = s^2).

The uncertainty $\sigma \simeq s$ calculated in Table 6-1 is indicated on the graph of Figure 6-1 by vertical error bars on each of the data points. The total height of each bar is equal to twice the P.E. = $\frac{2}{3}\sigma$ of Equation (3-11) to indicate the fact that the "true" value lies within $T_i \pm$ P.E. with a probability of 50%. The horizontal length of the ends of the bars is purely artistic, as is the size of the central dot. Notice that about half the data points miss the fitted line by more than the P.E.

Weighting with σ_i If the uncertainties are instrumental but not all equal, the calculations of the uncertainties in the parameters can be reduced equally simply. Equations (6-20) become, by taking the derivatives of Equations (6-12)

$$\frac{\partial a}{\partial y_j} = \frac{1}{\Delta} \left(\frac{1}{\sigma_j^2} \sum \frac{x_i^2}{\sigma_i^2} - \frac{x_j}{\sigma_j^2} \sum \frac{x_i}{x_j^2} \right)$$
$$\frac{\partial b}{\partial y_j} = \frac{1}{\Delta} \left(\frac{x_j}{\sigma_j^2} \sum \frac{1}{\sigma_i^2} - \frac{1}{\sigma_j^2} \sum \frac{x_i}{\sigma_j^2} \right) \qquad (6\text{-}24)$$

Combining these equations with the expression for the uncertainty in any coefficient in Equation (6-18), we have, following the procedure of Equations (6-21) and (6-22),

$$\sigma_a{}^2 \simeq \frac{1}{\Delta} \sum \frac{x_i{}^2}{\sigma_i{}^2}$$

$$\sigma_b{}^2 \simeq \frac{1}{\Delta} \sum \frac{1}{\sigma_i{}^2} \qquad\qquad (6\text{-}25)$$

$$\Delta = \sum \frac{1}{\sigma_i{}^2} \sum \frac{x_i{}^2}{\sigma_i{}^2} - \left(\sum \frac{x_i}{\sigma_i{}^2} \right)^2$$

In this case, the σ_i cannot be evaluated from the data and must be determined from a separate investigation of the experimental techniques. For example, if measurements are made with a non-linear scale meter, repeated observations y_i for one value of x_i will determine the absolute magnitude of the uncertainties, and the known non-linearity of the scale will yield the relative uncertainties.

Similarly, if the uncertainties are due to statistical fluctuations, we can substitute our approximation of Equation (6-16) for the uncertainties of the individual measurements (providing the y_i are raw data counts) into Equations (6-25).

$$\sigma_a{}^2 \simeq \frac{1}{\Delta} \sum \frac{x_i{}^2}{y_i}$$

$$\sigma_b{}^2 \simeq \frac{1}{\Delta} \sum \frac{1}{y_i} \qquad\qquad (6\text{-}26)$$

$$\Delta = \sum \frac{1}{y_i} \sum \frac{x_i{}^2}{y_i} - \left(\sum \frac{x_i}{y_i} \right)^2$$

The calculations of Equations (6-25) and (6-26) are illustrated in statements 67–68 of the subroutine LINFIT (VARNCE = 1. from statement 62 and is to be ignored). The method is the same for both, but the weighting factor WEIGHT $= 1/\sigma_i{}^2$, which appears in the accumulated sums, is calculated according to Equation (6-16) for statistical fluctuations (MODE < 0) and is calculated from the given values of SIGMAY(I) $= \sigma_I$ for instrumental uncertainties (MODE > 0).

The uncertainties σ_i of the data points for the experiment of Example 6-2 are indicated on the graph of Figure 6-2 by vertical error bars. The lengths are shown as twice the P.E. as for Example 6-1, even though the data are distributed according to the Poisson rather than the Gaussian distribution. Note, however, that the error bars decrease in length as the numbers of counts decrease according to Equation (6-16). As expected about half the data points miss the fitted line by more than the P.E.

Intuitive justification We can get a feeling for the magnitudes of the uncertainties in the coefficients σ_a and σ_b by considering the simplest case. Let us assume we have accumulated data points which have a common equal instrumental uncertainty $\sigma_i = \sigma$, and let us further suppose that the values of x are equally distributed about the origin so that $\Sigma x_i = 0$. (We can always redefine our origin so that this is true without affecting the magnitudes of the uncertainties.) According to Equations (6-21) and (6-22), we find that

$$\sigma_a = \frac{\sigma}{\sqrt{N}} \quad \text{and} \quad \sigma_b = \frac{\sigma}{\sqrt{\Sigma x_i^2}}$$

We can justify the expression for σ_a by considering the uncertainty in a to be similar to the uncertainty in the average of the data points $\overline{y_i} = (1/N)\Sigma y_i$. Our earlier rules for the propagation of errors indicate that the uncertainty in this average is reduced by $1/\sqrt{N}$ from the error in each point. The expression for σ_b is not so easy to justify intuitively, but it follows from the same reasoning, considering that b is the slope and is related to the ratio of y/x, whereas a is the intercept and is related to the absolute value of y. The magnitude of σ_b is therefore the same as that of σ_a, divided by the root mean square magnitude of the values of x

$$\sigma_b = \frac{\sigma/\sqrt{N}}{\sqrt{(1/N)\Sigma x_i^2}} = \frac{\sigma_a}{\sqrt{(1/N)\Sigma x_i^2}}$$

SUMMARY

Linear function:

$$y(x) = a + bx$$

Chi-square:

$$\chi^2 = \sum \left[\frac{1}{\sigma_i{}^2} (y_i - a - bx_i)^2 \right]$$

Least-squares fitting procedure: Minimize χ^2 with respect to each of the coefficients simultaneously.

Coefficients of least-squares fitting:

$$a = \frac{1}{\Delta} \left(\sum \frac{x_i{}^2}{\sigma_i{}^2} \sum \frac{y_i}{\sigma_i{}^2} - \sum \frac{x_i}{\sigma_i{}^2} \sum \frac{x_i y_i}{\sigma_i{}^2} \right)$$

$$b = \frac{1}{\Delta} \left(\sum \frac{1}{\sigma_i{}^2} \sum \frac{x_i y_i}{\sigma_i{}^2} - \sum \frac{x_i}{\sigma_i{}^2} \sum \frac{y_i}{\sigma_i{}^2} \right)$$

$$\Delta = \sum \frac{1}{\sigma_i{}^2} \sum \frac{x_i{}^2}{\sigma_i{}^2} - \left(\sum \frac{x_i}{\sigma_i{}^2} \right)^2$$

Estimated variance s^2:

$$\sigma^2 \simeq s^2 = \frac{1}{N-2} \Sigma(y_i - a - bx_i)^2$$

Statistical fluctuations:

$$\sigma_i{}^2 \simeq y_i \qquad \text{raw data counts}$$

Uncertainties in coefficients:

$$\sigma_a{}^2 \simeq \frac{1}{\Delta} \sum \frac{x_i{}^2}{\sigma_i{}^2} \qquad \sigma_b{}^2 \simeq \frac{1}{\Delta} \sum \frac{1}{\sigma_i{}^2}$$

EXERCISES

6-1 Fit the data of Example 6-2 as if all of the data had equal uncertainties $\sigma_i = \sigma$.

6-2 How would you go about solving the simultaneous equations of Equations (6-15)?

6-3 Fit the data of Example 6-1 as if all the uncertainties followed the Poisson distribution $\sigma_i{}^2 \simeq T_i$.

6-4 Derive Equations (6-25).

6-5 Compare the discrepancies Δ_i of Example 6-1 with the experimental uncertainty s. How much larger than s is the largest value of Δ_i? How probable is such a discrepancy?

6-6 Show that Equations (6-12) reduce to Equations (6-9) if $\sigma_i = \sigma$.

6-7 Derive a formula for making a linear fit to data with an intercept at the origin $y = bx$.

CHAPTER SEVEN

CORRELATION
PROBABILITY

7-1   LINEAR - CORRELATION COEFFICIENT

Let us assume that we have made measurements of pairs of
quantities x_i and y_i. We know from Chapter 6 how to make a least-
squares fit to these data for a linear relationship, and in the next
chapters we will consider fitting the data with more complex functions. But we must also stop and ask whether the fitting procedure
is justified, whether, indeed, there *exists* a physical relationship
between the variables x and y. What we are asking here is whether
or not the variations in the observed values of one quantity y are
correlated with the variations in the measured values of the other
quantity x.

For example, if we were to measure the length of a metal rod as a function of temperature, we would find a definite and reproducible correlation between the two quantities. But if we were to measure the length of the rod as a function of time, even though there might be fluctuations in the observations, we would not find any significant reproducible long-term relationship between the two sets of measurements.

On the basis of our discussion in Chapter 6, we can develop a quantitative measure of the degree of linear correlation or the probability that a linear relationship exists between two observed quantities. We can construct a linear-correlation coefficient r which will indicate quantitatively whether or not we are justified in determining even the simplest linear correspondence between the two quantities.

Reciprocity in fitting x vs. y Our data consist of pairs of measurements (x_i, y_i). If we consider the quantity y to be the dependent variable, then we want to know if the data correspond to a straight line of the form

$$y = a + bx \tag{7-1}$$

We have already developed the analytical solution for the coefficient b which represents the slope of the fitted line given in Equation (6-9),

$$b = \frac{N\Sigma x_i y_i - \Sigma x_i \Sigma y_i}{N\Sigma x_i{}^2 - (\Sigma x_i)^2} \tag{7-2}$$

where the weighting factors σ_i have been omitted for clarity. If there is no correlation between the quantities x and y, then there will be no tendency for the values of y to increase or decrease with increasing x, and, therefore, the least-squares fit must yield a horizontal straight line with a slope $b = 0$. But the value of b by itself cannot be a good measure of the degree of correlation since a relationship might exist which included a very small slope.

Since we are discussing the interrelationship between the variables x and y, we can equally well consider x as a function of y

and ask if the data correspond to a straight line of the form

$$x = a' + b'y \qquad (7\text{-}3)$$

The values of the coefficients a' and b' will be different from the values of the coefficients a and b in Equation (7-1), but they are related if the variables x and y are correlated.

The analytical solution for the inverse slope b' is similar to that for b in Equation (7-2).

$$b' = \frac{N\Sigma x_i y_i - \Sigma x_i \Sigma y_i}{N\Sigma y_i^2 - (\Sigma y_i)^2}$$

If there is no correlation between the quantities x and y, then the least-squares fit must yield a horizontal straight line with a slope $b' = 0$ as above for b.

If there is complete correlation between x and y, then there exists a relationship between the coefficients a and b of Equation (7-1) and between a' and b' of Equation (7-3). To see what this relationship is, we rewrite Equation (7-3)

$$y = -\frac{a'}{b'} + \frac{1}{b'} x = a + bx$$

and equate coefficients.

$$a = -\frac{a'}{b'}$$
$$b = \frac{1}{b'} \qquad (7\text{-}4)$$

If there is complete correlation, we see from Equation (7-4) that $bb' = 1$. If there is no correlation, both b and b' are 0. We therefore define the experimental linear-correlation coefficient $r \equiv \sqrt{bb'}$ as a measure of the degree of linear correlation.

$$r \equiv \frac{N\Sigma x_i y_i - \Sigma x_i \Sigma y_i}{[N\Sigma x_i^2 - (\Sigma x_i)^2]^{1/2}[N\Sigma y_i^2 - (\Sigma y_i)^2]^{1/2}} \qquad (7\text{-}5)$$

The value of r ranges from 0, when there is no correlation, to ± 1, when there is complete correlation. The sign of r is the same as that of b (and b'), but only the absolute magnitude is important.

The correlation coefficient r cannot be used directly to indicate the degree of correlation. A probability distribution for r can be derived from the two-dimensional Gaussian distribution, but its evaluation requires a knowledge of the correlation coefficient ρ of the parent population. A more common test of r is to compare its value with the probability distribution for a parent population which is completely uncorrelated, that is, for which $\rho = 0$. Such a comparison will indicate whether or not it is probable that the data points could represent a sample derived from an uncorrelated parent population. If this probability is small, then it is more probable that the data points represent a sample from a parent population where the variables are correlated.

For a parent population with $\rho = 0$, the probability that any random sample of uncorrelated experimental data points would yield an experimental linear-correlation coefficient equal to r is given by[1]

$$P_r(r,\nu) = \frac{1}{\sqrt{\pi}} \frac{\Gamma[(\nu + 1)/2]}{\Gamma(\nu/2)} (1 - r^2)^{(\nu-2)/2} \tag{7-6}$$

where $\nu = N - 2$ is the number of degrees of freedom for an experimental sample of N data points.

The gamma function $\Gamma(n)$ is equivalent to the factorial function $n!$ extended to nonintegral arguments. It is defined for integral and half-integral arguments by the values for arguments of 1 and $\frac{1}{2}$ and a recursion relation.

$$\Gamma(1) = 1 \qquad \Gamma(\tfrac{1}{2}) = \sqrt{\pi} \qquad \Gamma(n + 1) = n\Gamma(n)$$

For integral arguments

$$\Gamma(n + 1) = n! \qquad n = 0, 1, \ldots$$

For half-integral arguments

$$\Gamma(n + 1) = n(n - 1)(n - 2) \cdots (\tfrac{3}{2})(\tfrac{1}{2}\sqrt{\pi})$$
$$n = \tfrac{1}{2}, \tfrac{3}{2}, \tfrac{5}{2}, \ldots \tag{7-7}$$

[1] For a derivation see Pugh and Winslow, sec. 12-8.

Integral probability A more useful distribution than that of Equation (7-6) is the probability $P_c(r,N)$ that a random sample of N uncorrelated experimental data points would yield an experimental linear-correlation coefficient as large as or larger than the observed value of $|r|$. This probability is the integral of $P_r(r,\nu)$ for $\nu = N - 2$.

$$P_c(r,N) = 2 \int_{|r|}^{1} P_r(\rho,\nu)\, d\rho \qquad \nu = N - 2 \qquad (7\text{-}8)$$

With this definition, $P_c(r,N)$ indicates the probability that the observed data could have come from an uncorrelated ($\rho = 0$) parent population. A small value of $P_c(r,N)$ implies that the observed variables are probably correlated.

Program 7-1 The probability function $P_c(r,N)$ of Equation (7-8) can be computed by expanding the integral. For even values of ν the exponent is an integer and the binomial expansion can be used to expand the argument of the integral.

$$P_c(r,N) = \begin{cases} 1 - \dfrac{2}{\sqrt{\pi}} \dfrac{\Gamma[(\nu + 1)/2]}{\Gamma(\nu/2)} \\[2mm] \qquad \displaystyle\int_0^{|r|} \sum_{i=0}^{I} \left[(-1)^i \frac{I!}{(I-i)!\,i!} \rho^{2i} \right] d\rho \\[2mm] \qquad\qquad\qquad\qquad\qquad I = \tfrac{1}{2}(\nu - 2) \\[3mm] 1 - \dfrac{2}{\sqrt{\pi}} \dfrac{\Gamma[(\nu + 1)/2]}{\Gamma(\nu/2)} \\[2mm] \qquad \left\{ \displaystyle\sum_{i=0}^{I} \left[(-1)^i \frac{I!}{(I-i)!\,i!} \frac{|r|^{2i+1}}{2i+1} \right] \right\} \\[3mm] \qquad\qquad\qquad\qquad\qquad\qquad\qquad \nu \text{ even} \end{cases}$$

For odd values of ν, the exponent is half-integral and the expansion is more complex to derive, but the gamma functions may be included in the expansion to simplify the computation.

$$P_c(r,N) = 1 - \frac{1}{\sqrt{\pi}} \left\{ \sin^{-1}(|r|) \right.$$
$$\left. + |r| \sum_{i=\frac{1}{2}}^{I} \left[(1 - r^2)^i \frac{(2i-1)!!}{2i!!} \right] \right\} \qquad \nu \text{ odd}$$

Program 7-1 PCORRE Integral linear-correlation coefficient probability function $P_c(r,N)$.

```
C      FUNCTION PCORRE
C
C      PURPOSE
C        EVALUATE PROBABILITY FOR NO CORRELATION BETWEEN TWO VARIABLES
C
C      USAGE
C        RESULT = PCORRE (R, NPTS)
C
C      DESCRIPTION OF PARAMETERS
C        R       - LINEAR CORRELATION COEFFICIENT
C        NPTS    - NUMBER OF DATA POINTS
C
C      SUBROUTINES AND FUNCTION SUBPROGRAMS REQUIRED
C        GAMMA (X)
C          CALCULATES GAMMA FUNCTION FOR INTEGERS AND HALF-INTEGERS
C
C      MODIFICATIONS FOR FORTRAN II
C        OMIT DOUBLE PRECISION SPECIFICATIONS
C        ADD F SUFFIX TO ABS IN STATEMENTS 23 AND 42
C        CHANGE DSQRT TO SQRTF IN STATEMENT 42
C        CHANGE DATAN TO ATANF IN STATEMENT 43
C
```

The double factorial sign !! represents

$$n!! = n(n-2)(n-4) \cdots \begin{matrix} (3) & (1) & \text{for } n \text{ odd} \\ (4) & (2) & \text{for } n \text{ even} \end{matrix}$$

The computation of $P_c(r,N)$ is illustrated in the computer routine PCORRE of Program 7-1. This is a Fortran function subprogram to evaluate $P_c(r,N)$ for a given value of r and N. The input variables are R $= r$, the correlation coefficient to be tested, and NPTS $= N$, the number of data points.

FREE = NFREE $= \nu = N - 2$

is the number of degrees of freedom for a linear fit, and IMAX $= I$ is the number of terms in the expansion. The sum of terms is accumulated in statements 31–36 for ν even and in statements 51–56 for ν odd. The value of the probability is returned to the calling program as the value of the function PCORRE.

Program 7-2 The computer routine GAMMA of Program 7-2 is used to evaluate the gamma functions. Statements 11–13 determine whether the argument of the calling sequence X is integral or half-integral. If the argument is integral, the gamma function

```
      FUNCTION PCORRE (R, NPTS)
      DOUBLE PRECISION R2, TERM, SUM, FI, FNUM, DENOM
C
C         EVALUATE NUMBER OF DEGREES OF FREEDOM
C
   11 NFREE = NPTS - 2
      IF (NFREE) 13, 13, 15
   13 PCORRE = 0.
      GO TO 60
   15 R2 = R**2
      IF (1.-R2) 13, 13, 17
   17 NEVEN = 2*(NFREE/2)
      IF (NFREE - NEVEN) 21, 21, 41
C
C         NUMBER OF DEGREES OF FREEDOM IS EVEN
C
   21 IMAX = (NFREE-2)/2
      FREE = NFREE
   23 TERM = ABS (R)
      SUM = TERM
      IF (IMAX) 60, 26, 31
   26 PCORRE = 1. - TERM
      GO TO 60
   31 DO 36 I=1, IMAX
      FI = I
      FNUM = IMAX - I + 1
      DENOM = 2*I + 1
      TERM = -TERM * R2 * FNUM/FI
   36 SUM = SUM + TERM/DENOM
      PCORRE = 1.128379167 * (GAMMA((FREE+1.)/2.) / GAMMA(FREE/2.))
      PCORRE = 1. - PCORRE*SUM
      GO TO 60
C
C         NUMBER OF DEGREES OF FREEDOM IS ODD
C
   41 IMAX = (NFREE-3)/2
   42 TERM = ABS (R) * DSQRT(1.-R2)
   43 SUM = DATAN(R2/TERM)
      IF (IMAX) 57, 45, 51
   45 SUM = SUM + TERM
      GO TO 57
   51 SUM = SUM + TERM
   52 DO 56 I=1, IMAX
      FNUM = 2*I
      DENOM = 2*I + 1
      TERM = TERM * (1.-R2) * FNUM/DENOM
   56 SUM = SUM + TERM
   57 PCORRE = 1. - 0.6366197724*SUM
   60 RETURN
      END
```

is identical to the factorial function FACTOR(N) = $N!$ of Program 3-2, which is called in statement 21 to evaluate the result. If the argument is half-integral, the result GAMMA is set initially equal to $\Gamma(\frac{1}{2})$ in statement 31, and the product of Equations (7-7) is iterated in statements 41–43 for $x < 11$ and in statements 51–55 for $x > 10$.

Program 7-2 GAMMA Gamma function $\Gamma(n)$ for integers and half-integers.

```
C       FUNCTION GAMMA
C
C       PURPOSE
C         CALCULATE THE GAMMA FUNCTION FOR INTEGERS AND HALF-INTEGERS
C
C       USAGE
C         RESULT = GAMMA (X)
C
C       DESCRIPTION OF PARAMETERS
C         X       - INTEGER OR HALF-INTEGER
C
C       SUBROUTINES AND FUNCTION SUBPROGRAMS REQUIRED
C         FACTOR (N)
C             CALCULATES N FACTORIAL FOR INTEGERS
C
C       MODIFICATIONS FOR FORTRAN II
C         OMIT DOUBLE PRECISION SPECIFICATIONS
C         CHANGE DLOG TO LOGF IN STATEMENT 54
C         CHANGE DEXP TO EXPF IN STATEMENT 55
C
        FUNCTION GAMMA (X)
        DOUBLE PRECISION PROD, SUM, FI
C
C           INTEGERIZE ARGUMENT
C
   11 N = X - .25
      XN = N
   13 IF (X-XN-.75) 31, 31, 21
C
C           ARGUMENT IS INTEGER
C
   21 GAMMA = FACTOR(N)
      GO TO 60
C
C           ARGUMENT IS HALF-INTEGER
C
   31 PROD = 1.77245385
      IF (N) 44, 44, 33
   33 IF (N-10) 41, 41, 51
   41 DO 43 I=1, N
      FI = I
   43 PROD = PROD * (FI-.5)
   44 GAMMA = PROD
      GO TO 60
   51 SUM = 0.
      DO 54 I=11, N
      FI = I
   54 SUM = SUM + DLOG(FI-.5)
   55 GAMMA = PROD * 639383.8623 * DEXP(SUM)
   60 RETURN
      END
```

Sample calculation   The calculation of the linear-correlation coefficient R $= r$ is carried out in the subroutine LINFIT of Program 6-1. Statement 71 is equivalent to Equation (7-5) with provision for including the standard deviations σ_i of the data points as weighting factors. Note that SUM = $\Sigma(1/\sigma_i{}^2)$ is substituted for N = NPTS, and DELTA = Δ is substituted for the left-hand term in the denominator of Equation (7-5). Each of the sums includes the proper weighting by $\sigma_i{}^2$ as determined by the variable MODE (see discussion of Section 6-3).

EXAMPLE 7-1   In the experiment of Example 6-1, the linear-correlation coefficient r is given by Equation (7-5) to be

$$r = \frac{9(2898) - 45(466.7)}{\sqrt{9(285) - (45^2)} \sqrt{9(29,828.65) - (466.7)^2}}$$
$$= \frac{26,082 - 21,002}{\sqrt{540} \sqrt{50649}} = 0.97$$

From the graph of Figure C-3, a value of $r = 0.97$ with $N = 9$ observations yields a probability of determining such a large correlation from an uncorrelated population as $P_c(r,N) < 0.001$. This means that it is extremely improbable that the variables T and x are linearly uncorrelated; i.e., the probability is high that they are correlated and that our fit to a straight line is justified.

Similarly, in the experiment of Example 6-2, the linear-correlation coefficient is given by Equation (7-5) with the weighting factor $\sigma_i{}^2 = y_i$ introduced.

$$r = \frac{.1919(675) - 16.911(10)}{\sqrt{.1919(1837.74) - (16.911)^2} \sqrt{.1919(652) - (10)^2}}$$
$$= \frac{129.53 - 169.11}{\sqrt{66.66} \sqrt{25.119}} = -0.97$$

Again, the probability $P_c(r,N)$ for $r = 0.97$ with $N = 10$ observations is less than 0.001 indicating that the change in the counting rate C is linearly correlated with time t with a high degree of probability.

7-2 CORRELATION BETWEEN MANY VARIABLES

If the dependent variable y is a function of more than one variable,

$$y = a + b_1x_1 + b_2x_2 + b_3x_3 + \cdots \tag{7-9}$$

we might investigate the correlation between y and each of the independent variables x_j or we might also inquire into the possibility of correlation between the various different variables x_j if they are not independent.

To differentiate between the subscripts of Equations (7-5) and (7-9), let us use double subscripts on the variables x_{ij}. The first subscript i will represent the observation y_i as in the previous discussions. The second subscript j will represent the particular variable under investigation. Let us also rewrite Equation (7-5) for the linear-correlation coefficient r in terms of another quantity $s_{jk}{}^2$.

We define the *sample covariance* $s_{jk}{}^2$

$$s_{jk}{}^2 \equiv \frac{1}{N-1} \Sigma[(x_{ij} - \bar{x}_j)(x_{ik} - \bar{x}_k)] \tag{7-10}$$

where the means $\bar{x}_j$ and $\bar{x}_k$ are given by

$$\bar{x}_j \equiv \frac{1}{N} \Sigma x_{ij} \qquad \text{and} \qquad \bar{x}_k \equiv \frac{1}{N} \Sigma x_{ik} \tag{7-11}$$

and the sums are taken over the range of the subscript i from 1 to N. With this definition, the sample variance for one variable $s_j{}^2$

$$s_j{}^2 \equiv s_{jj}{}^2 = \frac{1}{N-1} \Sigma(x_{ij} - \bar{x}_j)^2 \tag{7-12}$$

is analogous to the sample variance $s_x{}^2$ defined in Equation (2-10).

$$s_x{}^2 = \frac{1}{N-1} \Sigma(x_i - \bar{x})^2$$

Equation (7-10) can be rewritten for comparison with Equation (7-5) by substituting the definitions of Equations (7-11).

$$s_{jk}{}^2 = \frac{1}{N-1} \Sigma(x_{ij}x_{ik} - \bar{x}_j\bar{x}_k)$$

$$= \frac{1}{N-1} (\Sigma x_{ij}x_{ik} - \frac{1}{N} \Sigma x_{ij}\Sigma x_{ik}) \tag{7-13}$$

If we substitute x_{ij} for x_i and x_{ik} for y_i in Equation (7-5), we can define the *sample linear-correlation coefficient* between any two variables x_j and x_k as

$$r_{jk} \equiv \frac{s_{jk}^2}{s_j s_k} \tag{7-14}$$

with the covariances and variances s_{jk}^2, s_j^2, and s_k^2 given by Equations (7-12) and (7-13). Thus, the linear-correlation coefficient between the jth variable x_j and the dependent variable y is given by

$$r_{jy} = \frac{s_{jy}^2}{s_j s_y} \tag{7-15}$$

Similarly, the linear-correlation coefficient of the parent population of which the data are a sample is defined as

$$\rho_{jk} = \frac{\sigma_{jk}^2}{\sigma_j \sigma_k}$$

where σ_j^2, σ_k^2, and σ_{jk}^2 are the true variances and covariances of the parent population. These linear-correlation coefficients are also known as product-moment-correlation coefficients.

With such a definition we can consider either the correlation between the dependent variable and any other variable r_{jy} or the correlation between any two variables r_{jk}. It is important to note, however, that the sample variances s_j^2 defined by Equation (7-12) are measures of the range of variation of the variables and not of the uncertainties as are the sample variances s^2 defined in Sections 5-2 and 6-5.

Polynomials   In Chapter 8 we will investigate functional relationships between y and x of the form

$$y = a + bx + cx^2 + dx^3 + \cdots \tag{7-16}$$

In a sense, this is a variation on the linear relationship of Equation (7-8) where the powers of the single independent variable x are considered to be various variables $x_j = x^j$. The correlation between the independent variable y and the mth term in the power series of Equation (7-16), therefore, can be expressed in terms of Equations (7-12)–(7-15).

$$r_{my} = \frac{s_{my}{}^2}{s_m s_y}$$

$$s_m{}^2 = \frac{1}{N-1}\left[\Sigma x_i{}^{2m} - \frac{1}{N}(\Sigma x_i{}^m)^2\right]$$

$$s_y{}^2 = \frac{1}{N-1}\left[\Sigma y_i{}^2 - \frac{1}{N}(\Sigma y_i)^2\right]$$

$$s_{my}{}^2 = \frac{1}{N-1}\left(\Sigma x_i{}^m y_i - \frac{1}{N}\Sigma x_i{}^m \Sigma y_i\right)$$

Weighted fit If the uncertainties of the data points are not all equal $\sigma_i \neq \sigma$, we must include the individual standard deviations σ_i as weighting factors in the definitions of variances, covariances, and correlation coefficients. From Section 6-3, the prescription for introducing weighting is to weight each term in a sum by the factor $1/\sigma_i{}^2$.

The formula for the correlation coefficient remains the same as Equations (7-14) and (7-15), but the formulas of Equations (7-10) and (7-12) for calculating the variances and covariances must be modified

$$
\begin{aligned}
s_{jk}{}^2 &\equiv \frac{\dfrac{1}{N-1}\sum\left[\dfrac{1}{\sigma_i{}^2}(x_{ij}-\bar{x}_j)(x_{ik}-\bar{x}_k)\right]}{\dfrac{1}{N}\sum\dfrac{1}{\sigma_i{}^2}} \\[2em]
s_j{}^2 &\equiv s_{jj}{}^2 = \frac{\dfrac{1}{N-1}\sum\left[\dfrac{1}{\sigma_i{}^2}(x_{ij}-\bar{x}_j)^2\right]}{\dfrac{1}{N}\sum\dfrac{1}{\sigma_i{}^2}}
\end{aligned}
\tag{7-17}
$$

where the means $\bar{x}_j$ and $\bar{x}_k$ are also weighted means.

$$\bar{x}_j \equiv \frac{\Sigma[(1/\sigma_i{}^2)x_{ij}]}{\Sigma(1/\sigma_i{}^2)}$$

Thus, the actual weighting factor is

$$\text{Weight}_i = \frac{1/\sigma_i^2}{(1/N)\Sigma(1/\sigma_i^2)}$$

as specified by the discussion of Chapter 4 and Section 10-1.

Multiple-correlation coefficient We can extrapolate the concept of the linear-correlation coefficient, which characterizes the correlation between two variables at a time, to include multiple correlations between groups of variables taken simultaneously.

The linear-correlation coefficient r of Equation (7-5) between y and x can be expressed in terms of the variances and covariances of Equations (7-17) and the slope b of a straight-line fit given in Equation (7-2).

$$r^2 = \frac{s_{xy}^4}{s_x^2 s_y^2} = b\frac{s_{xy}^2}{s_y^2}$$

In analogy with this definition of the linear-correlation coefficient, we define the *multiple-correlation coefficient* R to be the sum over similar terms for the variables of Equation (7-9).

$$R^2 \equiv \sum_{j=1}^n \left(b_j \frac{s_{jy}^2}{s_y^2}\right) = \sum_{j=1}^n \left(b_j \frac{s_j}{s_y} r_{jy}\right) \tag{7-18}$$

The linear-correlation coefficient r is useful for testing whether one particular variable should be included in the theoretical function to which the data are fit. The multiple-correlation coefficient R characterizes the fit of the data to the entire function. A comparison of the multiple-correlation coefficient for different functions is therefore useful in optimizing the theoretical functional form.

We will defer until Chapter 10 a discussion of how to use these correlation coefficients to determine the validity of including each term in the polynomial of Equation (7-16) or the arbitrary function of Equation (7-9).

SUMMARY

Function linear in coefficients:

$$y = a + \sum_{j=1}^{n} b_j x_j$$

Sample covariance s_{jk}^2:

$$s_{jk}^2 \equiv \frac{\dfrac{1}{N-1} \sum \left[\dfrac{1}{\sigma_i^2} (x_{ij} - \bar{x}_j)(x_{ik} - \bar{x}_k) \right]}{\dfrac{1}{N} \sum \dfrac{1}{\sigma_i^2}}$$

$$= \frac{1}{N-1} \left(\frac{\sum \dfrac{x_{ij} x_{ik}}{\sigma_i^2}}{\dfrac{1}{N} \sum \dfrac{1}{\sigma_i^2}} - N\bar{x}_j \bar{x}_k \right)$$

$$\bar{x}_j \equiv \frac{\Sigma(x_{ij}/\sigma_i^2)}{\Sigma(1/\sigma_i^2)}$$

Sample variance:

$$s_j^2 \equiv s_{jj}^2$$

Linear-correlation coefficient:

$$r_{jk} \equiv \frac{s_{jk}^2}{s_j s_k}$$

Probability $P_c(r,N)$ that any random sample of uncorrelated experimental data points would yield an experimental linear-correlation coefficient as large as or larger than $|r|$:

$$P_c(r, \nu + 2) = \int_{|r|}^{1} \frac{2}{\sqrt{\pi}} \frac{\Gamma[(\nu + 1)/2]}{\Gamma(\nu/2)} (1 - x^2)^{(\nu-2)/2} \, dx$$

Multiple-correlation coefficient R:

$$R^2 \equiv \sum_{j=1}^{n} \left(b_j \frac{s_{jy}^2}{s_y^2} \right) = \sum_{j=1}^{n} \left(b_j \frac{s_j}{s_y} r_{jy} \right)$$

EXERCISES

7-1 Find the linear-correlation coefficient r for Example 6-1.

7-2 If a set of data when fitted with Equation (7-1) yield a zero slope $b = 0$, what can you say about the linear-correlation coefficient r? Justify this value in terms of the correlation between x_i and y_i.

7-3 Find the linear-correlation coefficient r for Example 6-2.

7-4 Verify the expansion in the computation of $P_c(r,N)$.

7-5 Find the linear-correlation coefficient r_1 between x_i and y_i for the data of Example 8-1.

7-6 If the linear-correlation coefficient r_1 of Exercise 7-5 is computed by evaluating the slopes b and b' ($r_1 = \sqrt{bb'}$), should the slopes be computed by fitting the data with a linear polynomial or a quadratic polynomial?

7-7 Find the correlation coefficient r_2 between $x_i{}^2$ and y_i for the data of Example 8-1. Does the correlation justify the use of a quadratic term?

7-8 Express the multiple-correlation coefficient R in terms of x_{ij}, y_i, and their averages.

7-9 Evaluate the multiple-correlation coefficient R for the data of Example 8-1.

CHAPTER EIGHT

LEAST-SQUARES FIT
TO A POLYNOMIAL

8-1 ANALYTICAL METHOD

Suppose our data (x_i, y_i) were not fit well by a straight line. We might construct a more complex function which includes curvature and try varying the coefficients of this function to fit the data more closely. The most useful function for such a fit is a power-series polynomial

$$y = a + bx + cx^2 + dx^3 + \cdots \tag{8-1}$$

where the dependent variable y is expressed as a sum of a power series of the independent variable x with coefficients a, b, c, d, etc.

In Chapter 6 we introduced the method of least squares in

order to optimize the values of the coefficients a and b in our fit to a straight line. The same method applies to fitting a polynomial to data. Let us fit the data with a quadratic curve

$$y = a + bx + cx^2 \qquad (8\text{-}2)$$

to illustrate the method. We can then extrapolate the results to include polynomials of any order.

Instrumental uncertainties The method of least squares requires that we minimize χ^2, our measure of the goodness of fit to the data [see Equation (6-6)].

$$\chi^2 \equiv \sum \left(\frac{\Delta y_i}{\sigma_i}\right)^2 = \sum \left[\frac{1}{\sigma_i^2}(y_i - a - bx_i - cx_i^2)^2\right] \qquad (8\text{-}3)$$

The minimum value of χ^2 can be determined by setting the derivatives of χ^2 with respect to each of the coefficients equal to 0. See the discussion of Appendix A.

Setting the derivatives of χ^2 in Equation (8-3) with respect to each of the three coefficients a, b, and c equal to 0 yields three simultaneous equations

$$\frac{\partial}{\partial a}\chi^2 = -2\sum\left[\frac{1}{\sigma_i^2}(y_i - a - bx_i - cx_i^2)\right] = 0$$

$$\frac{\partial}{\partial b}\chi^2 = -2\sum\left[\frac{x_i}{\sigma_i^2}(y_i - a - bx_i - cx_i^2)\right] = 0$$

$$\frac{\partial}{\partial c}\chi^2 = -2\sum\left[\frac{x_i^2}{\sigma_i^2}(y_i - a - bx_i - cx_i^2)\right] = 0$$

similar to the two simultaneous equations of Equations (6-10).

These can be rearranged to show the interaction of the coefficients explicitly

$$
\begin{aligned}
\Sigma y_i &= a\Sigma 1 &&+ b\Sigma x_i &&+ c\Sigma x_i^2 \\
\Sigma x_i y_i &= a\Sigma x_i &&+ b\Sigma x_i^2 &&+ c\Sigma x_i^3 \\
\Sigma x_i^2 y_i &= a\Sigma x_i^2 &&+ b\Sigma x_i^3 &&+ c\Sigma x_i^4
\end{aligned}
\qquad (8\text{-}4)
$$

where we have omitted a factor of σ_i^2 in the denominator of each term inside each summation sign to show the pattern more clearly.

Using the same determinant methods for solving these simultaneous equations that we introduced in Chapter 6 (see Appendix B), we can write the optimum values for the coefficients in terms of determinants.

$$a = \frac{1}{\Delta} \begin{vmatrix} \sum \dfrac{y_i}{\sigma_i{}^2} & \sum \dfrac{x_i}{\sigma_i{}^2} & \sum \dfrac{x_i{}^2}{\sigma_i{}^2} \\[2mm] \sum \dfrac{x_i y_i}{\sigma_i{}^2} & \sum \dfrac{x_i{}^2}{\sigma_i{}^2} & \sum \dfrac{x_i{}^3}{\sigma_i{}^2} \\[2mm] \sum \dfrac{x_i{}^2 y_i}{\sigma_i{}^2} & \sum \dfrac{x_i{}^3}{\sigma_i{}^2} & \sum \dfrac{x_i{}^4}{\sigma_i{}^2} \end{vmatrix}$$

$$b = \frac{1}{\Delta} \begin{vmatrix} \sum \dfrac{1}{\sigma_i{}^2} & \sum \dfrac{y_i}{\sigma_i{}^2} & \sum \dfrac{x_i{}^2}{\sigma_i{}^2} \\[2mm] \sum \dfrac{x_i}{\sigma_i{}^2} & \sum \dfrac{x_i y_i}{\sigma_i{}^2} & \sum \dfrac{x_i{}^3}{\sigma_i{}^2} \\[2mm] \sum \dfrac{x_i{}^2}{\sigma_i{}^2} & \sum \dfrac{x_i{}^2 y_i}{\sigma_i{}^2} & \sum \dfrac{x_i{}^4}{\sigma_i{}^2} \end{vmatrix}$$

(8-5)

$$c = \frac{1}{\Delta} \begin{vmatrix} \sum \dfrac{1}{\sigma_i{}^2} & \sum \dfrac{x_i}{\sigma_i{}^2} & \sum \dfrac{y_i}{\sigma_i{}^2} \\[2mm] \sum \dfrac{x_i}{\sigma_i{}^2} & \sum \dfrac{x_i{}^2}{\sigma_i{}^2} & \sum \dfrac{x_i y_i}{\sigma_i{}^2} \\[2mm] \sum \dfrac{x_i{}^2}{\sigma_i{}^2} & \sum \dfrac{x_i{}^3}{\sigma_i{}^2} & \sum \dfrac{x_i{}^2 y_i}{\sigma_i{}^2} \end{vmatrix}$$

$$\Delta = \begin{vmatrix} \sum \dfrac{1}{\sigma_i{}^2} & \sum \dfrac{x_i}{\sigma_i{}^2} & \sum \dfrac{x_i{}^2}{\sigma_i{}^2} \\[2mm] \sum \dfrac{x_i}{\sigma_i{}^2} & \sum \dfrac{x_i{}^2}{\sigma_i{}^2} & \sum \dfrac{x_i{}^3}{\sigma_i{}^2} \\[2mm] \sum \dfrac{x_i{}^2}{\sigma_i{}^2} & \sum \dfrac{x_i{}^3}{\sigma_i{}^2} & \sum \dfrac{x_i{}^4}{\sigma_i{}^2} \end{vmatrix}$$

Statistical fluctuations   If the errors are statistical errors, we may replace the weighting factors of $\sigma_i{}^2$ in the denominators with the approximation $\sigma_i{}^2 \simeq y_i$ as discussed in Section 6-4. If the errors are due to instrumental uncertainties such that all the individual standard deviations σ_i pertaining to our measurements y_i are equal to the same value σ, we can ignore those factors which we omitted from Equations (8-4), and Equations (8-5) reduce to (with the substitution of N for $\Sigma 1$)

$$a = \frac{1}{\Delta} \begin{vmatrix} \Sigma y_i & \Sigma x_i & \Sigma x_i^2 \\ \Sigma x_i y_i & \Sigma x_i^2 & \Sigma x_i^3 \\ \Sigma x_i^2 y_i & \Sigma x_i^3 & \Sigma x_i^4 \end{vmatrix} \qquad b = \frac{1}{\Delta} \begin{vmatrix} N & \Sigma y_i & \Sigma x_i^2 \\ \Sigma x_i & \Sigma x_i y_i & \Sigma x_i^3 \\ \Sigma x_i^2 & \Sigma x_i^2 y_i & \Sigma x_i^4 \end{vmatrix}$$

$$c = \frac{1}{\Delta} \begin{vmatrix} N & \Sigma x_i & \Sigma y_i \\ \Sigma x_i & \Sigma x_i^2 & \Sigma x_i y_i \\ \Sigma x_i^2 & \Sigma x_i^3 & \Sigma x_i^2 y_i \end{vmatrix} \qquad \Delta = \begin{vmatrix} N & \Sigma x_i & \Sigma x_i^2 \\ \Sigma x_i & \Sigma x_i^2 & \Sigma x_i^3 \\ \Sigma x_i^2 & \Sigma x_i^3 & \Sigma x_i^4 \end{vmatrix}$$

$$\tag{8-6}$$

The experimental variance s^2 is defined in the same way as the experimental variance for a straight-line fit of Equation (6-19)

$$s^2 = \frac{1}{N - n - 1} \Sigma (y_i - a - bx_i - cx_i^2)^2 \tag{8-7}$$

where $N - n - 1$ is the number of degrees of freedom ν for a fit of a polynomial of degree n to a set of N data points (for a quadratic $n = 2$).

Polynomials of degree n The extrapolation of these techniques to higher-order polynomials is straightforward. If we consider an nth degree polynomial,

$$y = a + bx + cx^2 + \cdots + mx^n$$

then Equations (8-4) expand to $n + 1$ simultaneous equations in $n + 1$ coefficients. The terms on the left sides of the equations range from Σy_i to $\Sigma x_i^n y_i$, and the terms on the right sides range from $\Sigma 1$ to Σx_i^{2n}. The determinants in Equations (8-5) and (8-6) expand to order $n + 1$. The extrapolation of the determinant for Δ to higher order is obvious. The determinants for the coefficients are identical in all but one row to that for Δ so that the extrapolation is similar.

EXAMPLE 8-1 A thermocouple is to be used for monitoring temperatures and we need to calibrate it against a thermometer. The thermocouple consists of a junction of a copper wire and a wire of constantan. In order to measure the junction voltage with high precision, we connect the sample junction in series with a reference junction which is held at room temperature. Our data, therefore, will be valid only for calibrating the relative variation of junction voltage with temperature. The absolute voltage must be determined in a separate experiment by measuring it at one specific temperature.

The difference in output voltage between the two junctions is measured for a variation in temperature of the sample junction from 0° to 100°C in steps of 5°C. The measurements are made on two scales of the voltmeter, the 1 mV scale and the 3 mV scale, but fluctuations of the needle on the 1 mV scale preclude the possibility of utilizing the inherent precision of the meter to reduce the uncertainty on this scale. Thus, we assume that the uncertainties in the measurements are equal throughout, and that they are estimated to be on the order of 0.05 mV.

The data of this experiment are given in Table 8-1 and

Table 8-1 Experimental data for a determination of the relative output voltage of a thermocouple junction V as a function of temperature T of the junction. Components of the junction are copper and constantan; the reference junction is at room temperature

Trial i	Temperature T, °C	Voltage V, mV	Fit $V(T_i)$	Trial i	Temperature T, °C	Voltage V, mV	Fit $V(T_i)$
1	0.	--0.89	−0.886	12	55.	1.22	1.233
2	5.	−0.69	−0.708	13	60.	1.45	1.438
3	10.	−0.53	−0.528	14	65.	1.68	1.657
4	15.	−0.34	−0.344	15	70.	1.88	1.874
5	20.	−0.15	−0.158	16	75.	2.10	2.093
6	25.	0.02	0.032	17	80.	2.31	2.316
7	30.	0.20	0.225	18	85.	2.54	2.541
8	35.	0.42	0.421	19	90.	2.78	2.770
9	40.	0.61	0.619	20	95.	3.00	3.001
10	45.	0.82	0.821	21	100.	3.22	3.236
11	50.	1.03	1.025				

$V = -0.8862 \text{ mV} + 0.03524T \text{ mV/°C} + 5.979 \times 10^{-5}T^2 \text{ mV/(°C)}^2$
$s = 0.01 \text{ mV}$

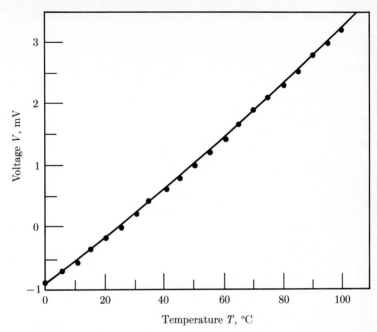

FIGURE 8-1 Graph of output voltage from thermocouple junction V relative to junction at room temperature vs. temperature T of thermocouple junction. Probable error for data points is estimated at 0.01 mV. Quadratic-polynomial fit is for $V = -0.886$ mV $+ 0.0352T$ mV/°C $+ 5.98 \times 10^{-5}T^2$ mV/(°C)^2.

plotted in Figure 8-1. To a first approximation, the variation is linear but there is a noticeable curvature. Theoretically, we expect these data to be fit well with a quadratic curve. The coefficients given in Table 8-1 are calculated according to Equations (8-6). The resultant fit is shown in Figure 8-1 as the solid line.

Note that although the curvature is slight, and the uncertainties in the individual data points are not small, there are enough data points to determine the coefficients fairly precisely. We have not used our estimated error for the data points of 0.05 mV for the calculation of the coefficients; we need only know that the standard deviations were all equal. In fact, the uncertainty

Program 8-1   POLFIT   Least-squares fit to a polynomial.

```
C     SUBROUTINE POLFIT
C
C     PURPOSE
C       MAKE A LEAST-SQUARES FIT TO DATA WITH A POLYNOMIAL CURVE
C         Y = A(1) + A(2)*X + A(3)*X**2 + A(4)*X**3 + . . .
C
C     USAGE
C       CALL POLFIT (X, Y, SIGMAY, NPTS, NTERMS, MODE, A, CHISQR)
C
C     DESCRIPTION OF PARAMETERS
C       X       - ARRAY OF DATA POINTS FOR INDEPENDENT VARIABLE
C       Y       - ARRAY OF DATA POINTS FOR DEPENDENT VARIABLE
C       SIGMAY  - ARRAY OF STANDARD DEVIATIONS FOR Y DATA POINTS
C       NPTS    - NUMBER OF PAIRS OF DATA POINTS
C       NTERMS  - NUMBER OF COEFFICIENTS (DEGREE OF POLYNOMIAL + 1)
C       MODE    - DETERMINES METHOD OF WEIGHTING LEAST-SQUARES FIT
C                 +1 (INSTRUMENTAL) WEIGHT(I) = 1./SIGMAY(I)**2
C                  0 (NO WEIGHTING) WEIGHT(I) = 1.
C                 -1 (STATISTICAL)  WEIGHT(I) = 1./Y(I)
C       A       - ARRAY OF COEFFICIENTS OF POLYNOMIAL
C       CHISQR  - REDUCED CHI SQUARE FOR FIT
C
C     SUBROUTINES AND FUNCTION SUBPROGRAMS REQUIRED
C       DETERM (ARRAY, NORDER)
C         EVALUATES THE DETERMINANT OF A SYMMETRIC TWO-DIMENSIONAL
C         MATRIX OF ORDER NORDER
C
C     MODIFICATIONS FOR FORTRAN II
C       OMIT DOUBLE PRECISION SPECIFICATIONS
C
C     COMMENTS
C       DIMENSION STATEMENT VALID FOR NTERMS UP TO 10
C
```

$s = 0.01$ mV estimated from the data is much smaller. If we calculate the experimental variance s^2 according to Equation (8-7), however, we will not be able to determine χ^2 since they are related:

$$s^2 = \frac{1}{N - n - 1} \chi^2 = \text{reduced } \chi^2$$

where $N - n - 1$ is the number of degrees of freedom for a fit of a polynomial of degree n to N data points. We will defer our discussion of the usefulness of χ^2 in determining the uncertainties in our values of the coefficients until Chapter 10.

Program 8-1   The calculations developed in this section are illustrated with the computer routine POLFIT of Program 8-1. This is a Fortran subroutine to calculate the coefficients a, b, c, d, etc. for a least-squares fit to an array of data points with the polynomial curve given in Equation (8-1). The calling sequence is

```
        SUBROUTINE POLFIT (X, Y, SIGMAY, NPTS, NTERMS, MODE, A, CHISQR)
        DOUBLE PRECISION SUMX, SUMY, XTERM, YTERM, ARRAY, CHISQ
        DIMENSION X(1), Y(1), SIGMAY(1), A(1)
        DIMENSION SUMX(19), SUMY(10), ARRAY(10,10)
C
C          ACCUMULATE WEIGHTED SUMS
C
   11 NMAX = 2*NTERMS - 1
        DO 13 N=1, NMAX
   13 SUMX(N) = 0.
        DO 15 J=1, NTERMS
   15 SUMY(J) = 0.
        CHISQ = 0.
   21 DO 50 I=1, NPTS
        XI = X(I)
        YI = Y(I)
   31 IF (MODE) 32, 37, 39
   32 IF (YI) 35, 37, 33
   33 WEIGHT = 1. / YI
        GO TO 41
   35 WEIGHT = 1. / (-YI)
        GO TO 41
   37 WEIGHT = 1.
        GO TO 41
   39 WEIGHT = 1. / SIGMAY(I)**2
   41 XTERM = WEIGHT
        DO 44 N=1, NMAX
        SUMX(N) = SUMX(N) + XTERM
   44 XTERM = XTERM * XI
   45 YTERM = WEIGHT*YI
        DO 48 N=1, NTERMS
        SUMY(N) = SUMY(N) + YTERM
   48 YTERM = YTERM * XI
   49 CHISQ = CHISQ + WEIGHT*YI**2
   50 CONTINUE
C
C          CONSTRUCT MATRICES AND CALCULATE COEFFICIENTS
C
   51 DO 54 J=1, NTERMS
        DO 54 K=1, NTERMS
        N = J + K - 1
   54 ARRAY(J,K) = SUMX(N)
        DELTA = DETERM (ARRAY, NTERMS)
        IF (DELTA) 61, 57, 61
   57 CHISQR = 0.
        DO 59 J=1, NTERMS
   59 A(J) = 0.
        GO TO 80
   61 DO 70 L=1, NTERMS
   62 DO 66 J=1, NTERMS
        DO 65 K=1, NTERMS
        N = J + K - 1.
   65 ARRAY(J,K) = SUMX(N)
   66 ARRAY(J,L) = SUMY(J)
   70 A(L) = DETERM (ARRAY, NTERMS) / DELTA
```

similar to that of the subroutine LINFIT in Program 6-1. The input
variables are X, Y, SIGMAY, NPTS, NTERMS, and MODE, and the output variables are A and CHISQR.

Program 8-1 POLFIT (*continued*)

```
C
C        CALCULATE CHI SQUARE
C
   71 DO 75 J=1, NTERMS
      CHISQ = CHISQ - 2.*A(J)*SUMY(J)
      DO 75 K=1, NTERMS
      N = J + K - 1
   75 CHISQ = CHISQ + A(J)*A(K)*SUMX(N)
   76 FREE = NPTS - NTERMS
   77 CHISQR = CHISQ / FREE
   80 RETURN
      END
```

The variable NPTS represents the number of pairs of data points NPTS $= N$. The variable NTERMS represents the degree of the polynomial desired; NTERMS is equal to the number of terms in the power series (i.e., the number of coefficients) which is one more than the degree n of the polynomial NTERMS $= n + 1$. For example, for the quadratic function of degree 2 in Equation (8-2), there are three coefficients, a, b, and c, and NTERMS $= 3$. The independent quantities x_i are stored in the array X, and the dependent data points are stored in the array Y, with the ordering identical for the two arrays.

The array SIGMAY contains the standard deviations σ_i for the measurements y_i, with the same ordering as for the data in arrays X and Y. The variable MODE determines the method of calculation of the weighting factor $1/\sigma_i^2$ in Equations (8-5) similar to its use in the subroutine LINFIT. If MODE $= 0$, the calculation proceeds on the assumption that the standard deviations are all equal $\sigma_i = \sigma$ and the weighting factor is ignored (set $= 1.$ in statement 37). If MODE $= +1$ (or any other positive nonzero integer), the subroutine uses the standard deviations stored in the array SIGMAY to calculate the weighting factor in statement 39. If MODE $= -1$ (or any other negative integer), the calculation proceeds on the assumption that the uncertainties are statistical and uses the approximation of Equation (6-16) $\sigma_i^2 \simeq y_i$ to calculate the weighting factor in statement 33 (or statement 35 to provide for negative values of Y). Note that SIGMAY is ignored if MODE is not positive. In this case the name of any array (for example, X or Y) may be substituted for SIGMAY in the calling sequence because the array is neither used nor altered.

The method of calculation is to construct the matrices of Equations (8-5) and to evaluate their determinants. Consider the matrix of which Δ is the determinant. The contents of any element in the matrix depend only on the distance of that element from the first element in the upper-left corner, counting only horizontally and/or vertically. The exponent appearing in any element which is m units away from the first element is m. For example, the third element along the top row is the same as the third element down the first column, and these are the same as the middle element. Each of these elements is two units away from the first element, and therefore the exponent of x_i is equal to 2.

If the degree of the polynomial is n, there are $n + 1$ rows and columns and $(n + 1)^2$ elements, but only $2n + 1$ different elements. These elements are calculated in statements 41–44 and stored in the linear array SUMX. Similarly, the $n + 1$ elements which make up the column that distinguishes the matrices for the coefficients from that of Δ are calculated in statements 45–48 and stored in the linear array SUMY. Since these elements are actually sums of terms, their accumulation is carried out in the DO loop extending from statement 21 to statement 50.

The matrix corresponding to Δ is constructed in statements 51–54, and in the following statement the variable DELTA = Δ is set equal to its determinant. In statements 61–70, the $n + 1$ other matrices are constructed from this matrix one by one. In statements 62–65, the matrix corresponding to Δ is constructed and stored in the array ARRAY. In statement 66, the Lth column is replaced by the linear array SUMY. The Lth coefficient A(L) = a_{L+1} is calculated in statement 70.

The rest of the subroutine pertains to a calculation of χ^2 and will be discussed in Chapter 10. The subroutine DETERM which evaluates the determinant of a matrix is explained in Appendix B.

8-2 INDEPENDENT PARAMETERS

Suppose we were to take the data of Example 6-1 or Example 6-2 and fit them with the quadratic polynomial function of Equation (8-2). We would expect to find a rather small and possibly meaningless result for the coefficient of the quadratic term c. But the fact that c is not set equal to 0 by definition, as in the analysis of Chapter 6, means that we may also find considerably different values for the coefficients a and b of our fit. In general, the polynomial fitting procedure of Section 8-1 will yield values for the coefficients which are dependent on the degree of the polynomial with which the fit is made.

This interdependence arises from the fact that we have specified our coordinate system without regard to the region of coefficient space from which our data points are extracted. The value of a represents the intercept of the curve on the ordinal axis. The coefficient b represents the slope of the curve at this same point, and the other coefficients represent higher orders of curvature at the same intercept point. If the data are not clustered about this intercept point, its location might be highly dependent on the degree of the polynomial used to fit the data.

We might be able to extract more meaningful information about the data if we were to determine instead coefficients a', b', c', etc., which represented the average value, the average slope, the average curvature, etc., of the data. Such coefficients would be independent of our choice of coordinate system and would represent physical characteristics of the data, which are independent of the degree of the polynomial being fit.

Orthogonal polynomials We would like to fit the data to a function which is similar to that of Equation (8-1) but which yields this desired independence of coefficients. The appropriate function to use is a sum of orthogonal polynomials[1] which has the

---

[1] Any polynomial such as that of Equation (8-1) can be rewritten as a sum of orthogonal polynomials

$$y = a + \sum_{j=1}^{n} [b_j X_j(x_i)]$$

with the orthogonal property that

$$\Sigma[X_j(x_i)X_k(x_i)] = 0 \qquad \text{for } j \neq k$$

form

$$y(x) = a + b(x - \beta) + c(x - \gamma_1)(x - \gamma_2)$$
$$+ d(x - \delta_1)(x - \delta_2)(x - \delta_3) + \cdots$$

Following the development of Section 8-1, we must minimize χ^2 to determine the coefficients a, b, c, d, etc., with the further criterion that the addition of higher-order terms to the polynomial will not affect the evaluation of lower-order terms. This criterion will be used to determine the parameters β, γ_1, γ_2, etc.

χ^2 is defined as

$$\chi^2 \equiv \sum \left(\frac{\Delta y_i}{\sigma_i}\right)^2 = \sum \left\{\frac{1}{\sigma_i^2}[y_i - y(x_i)]^2\right\}$$

Setting the derivatives of χ^2 with respect to each of the coefficients equal to 0 yields $n + 1$ simultaneous equations

$$\Sigma y_i = Na + b\Sigma(x_i - \beta) + c\Sigma(x_i - \gamma_1)(x_i - \gamma_2)$$
$$+ d\Sigma(x_i - \delta_1)(x_i - \delta_2)(x_i - \delta_3) + \cdots \quad (8\text{-}8)$$
$$\Sigma x_i y_i = a\Sigma x_i + b\Sigma x_i(x_i - \beta) + c\Sigma x_i(x_i - \gamma_1)(x_i - \gamma_2)$$
$$+ d\Sigma x_i(x_i - \delta_1)(x_i - \delta_2)(x_i - \delta_3) + \cdots \quad (8\text{-}9)$$
$$\Sigma x_i^2 y_i = a\Sigma x_i^2 + b\Sigma x_i^2(x_i - \beta) + c\Sigma x_i^2(x_i - \gamma_1)(x_i - \gamma_2)$$
$$+ d\Sigma x_i^2(x_i - \delta_1)(x_i - \delta_2)(x_i - \delta_3) + \cdots \quad (8\text{-}10)$$
$$\Sigma x_i^3 y_i = a\Sigma x_i^3 + b\Sigma x_i^3(x_i - \beta) + c\Sigma x_i^3(x_i - \gamma_1)(x_i - \gamma_2)$$
$$+ d\Sigma x_i^3(x_i - \delta_1)(x_i - \delta_2)(x_i - \delta_3) + \cdots \quad (8\text{-}11)$$

where, again, we have omitted a factor of σ_i^2 in the denominator of each term for clarity.

Parameters Let us examine Equation (8-8). If we restrict ourselves to a 0th degree polynomial, there is only one coefficient a; all the other coefficients are set equal to 0 by definition. The coefficient a, therefore, is specified completely by Equation (8-8), neglecting all but the first term on the right-hand side.

$$a = \frac{1}{N} \Sigma y_i = \bar{y} \tag{8-12}$$

If we restrict ourselves to a first-degree polynomial, the coefficient b of the second term of Equation (8-8) is not 0. If we are to have independence of coefficients, however, this term must still be 0. Hence, the conclusion that the sum in this term is 0

$$\Sigma(x_i - \beta) = 0$$

leads to a value for β.

$$\beta = \frac{1}{N} \Sigma x_i = \bar{x} \tag{8-13}$$

Similarly, if we consider a quadratic function, the third term of Equation (8-8) must be 0 even when the coefficient c is not 0. This constraint leads to a quadratic equation in γ_1 and γ_2 which is not sufficient to specify either parameter. We have the additional constraint, however, that the coefficient b must be specified completely by Equations (8-8) and (8-9); that is, Equation (8-9) must determine b after a is determined by Equation (8-8). Thus, the third term in Equation (8-9) must also be 0 regardless of the value of the coefficient c, and we have two simultaneous quadratic equations for the parameters γ_1 and γ_2.

$$\begin{aligned}
\Sigma(x_i - \gamma_1)(x_i - \gamma_2) &= 0 \\
\Sigma x_i(x_i - \gamma_1)(x_i - \gamma_2) &= 0
\end{aligned} \tag{8-14}$$

Similarly, the coefficient c must be determined completely by Equation (8-10) (and the predetermined values of a and b), and this constraint yields three simultaneous equations for the parameters δ_1, δ_2, and δ_3.

$$\begin{aligned}
\Sigma(x_i - \delta_1)(x_i - \delta_2)(x_i - \delta_3) &= 0 \\
\Sigma x_i(x_i - \delta_1)(x_i - \delta_2)(x_i - \delta_3) &= 0 \\
\Sigma x_i^2(x_i - \delta_1)(x_i - \delta_2)(x_i - \delta_3) &= 0
\end{aligned} \tag{8-15}$$

The extrapolation to higher-order parameters is straightforward.

Coefficients   Once the parameters β, γ, δ, etc., are determined by the constraints described above, the coefficients a, b, c, etc., can be determined from the resulting $n + 1$ simultaneous equations. The value for the first coefficient a is specified completely by minimizing χ^2 with respect to a in Equation (8-8) and is given in Equation (8-12). The value of the second coefficient b is determined by minimizing χ^2 with respect to both a and b in Equations (8-8) and (8-9). Substituting the value of a from Equation (8-12) into Equation (8-9) yields a result for b directly. Similarly, the value for c can be determined from Equation (8-10) after substituting the values of a and b determined from Equations (8-8) and (8-9). Each succeeding equation yields the value for the next higher-order coefficient.

Note that the value of any coefficient is therefore independent of the value specified for any higher-order coefficient but is not independent of the values of lower-order coefficients. Thus, the coefficients obtained are given by

$$a = \frac{1}{N} \Sigma y_i = \bar{y}$$

$$b = \frac{\Sigma y_i(x_i - \beta)}{\Sigma(x_i - \beta)^2}$$

$$c = \frac{\Sigma[y_i(x_i - \gamma_1)(x_i - \gamma_2)]}{\Sigma[(x_i - \gamma_1)(x_i - \gamma_2)]^2}$$

$$d = \frac{\Sigma[y_i(x_i - \delta_1)(x_i - \delta_2)(x_i - \delta_3)]}{\Sigma[(x_i - \delta_1)(x_i - \delta_2)(x_i - \delta_3)]^2}$$

and so on.

Simplification For the general case of arbitrarily chosen sets of data points (x_i, y_i), this procedure is cumbersome even with computer techniques. There is, however, a special type of data for which the calculations can be considerably simplified, namely, whenever the values of the independent quantity x are equally spaced and when we can consider all the uncertainties σ_i equal and ignore them as we have in the development above.

Consider the experiments of Examples 6-1 and 8-1. Their data fulfill these two criteria, and, therefore, we could use this method to fit the data. The values of the coefficients for these experiments might not have any greater physical significance (that is, $a = \bar{T}$ the average temperature for the data points is not a fundamental constant), but by using orthogonal polynomials we could try fitting the data with higher-degree polynomials without changing the values of the coefficients already calculated for a straight-line or quadratic fit. Similarly, the experiment of Example 6-2 fulfills the first criterion, that the x data points are equally spaced throughout the experiment. However, the uncertainties of Example 6-2 are statistical, and we cannot ignore the factor of $\sigma_i{}^2$ in the denominators of Equations (8-8) to (8-11).

For an experiment similar to that of Example 8-1, where we have made N measurements for equally spaced values of the independent variable x ranging from x_1 to x_N in steps of Δ,

$$\Delta = x_{i+1} - x_i$$

and assuming that the uncertainties are due to instrumental errors with a common standard deviation $\sigma_i = \sigma$, Equations (8-13) to (8-15) reduce to

$$\beta = \frac{1}{N} \Sigma x_i = \bar{x} = \frac{1}{2} (x_1 + x_N)$$

$$\gamma = \beta \pm \sqrt{\frac{1}{N} \Sigma (x_i - \beta)^2} = \beta \pm \Delta \sqrt{\frac{1}{12} (N^2 - 1)}$$

$$\delta = \beta, \beta \pm \sqrt{\frac{\Sigma[x_i(x_i - \beta)^3]}{\Sigma[x_i(x_i - \beta)]}} = \beta, \beta \pm \Delta \sqrt{\frac{1}{20} (3N^2 - 7)}$$

A more comprehensive list of parameters for orthogonal polynomials can be found in Anderson and Houseman.

8-3 MATRIX INVERSION

The method of least squares developed in Section 8-1 can be adapted equally well to fitting data with functions other than the power-series polynomial given in Equation (8-1). It is sufficient that the function be linear in each of the coefficients (for example, $y = ae^{bx}$ is linear in a but not in b).

Legendre polynomials   One important function for which the method is applicable is the *Legendre polynomial*

$$y = a_0 P_0(x) + a_1 P_1(x) + \cdots = \sum_{L=0}^{n} [a_L P_L(x)] \qquad (8\text{-}16)$$

where $x = \cos \theta$ and the terms $P_L(x)$ in the function are given by

$$P_0(x) = 1 \qquad P_2(x) = \tfrac{1}{2}(3x^2 - 1)$$
$$P_1(x) = x \qquad P_3(x) = \tfrac{1}{2}(5x^3 - 3x) \qquad (8\text{-}17)$$

and higher-order terms can be determined from the iteration formula.

$$P_L(x) = \frac{1}{L}[(2L - 1)xP_{L-1}(x) - (L - 1)P_{L-2}(x)] \qquad (8\text{-}18)$$

Legendre polynomials are orthogonal when averaged over all values of $x = \cos \theta$

$$\int_{-1}^{1} [P_L(x)P_M(x)]\, dx = \begin{cases} 1 & \text{if } L = M \\ 0 & \text{if } L \neq M \end{cases}$$

EXAMPLE 8-2 When ^{13}C is bombarded by protons with 4.5 MeV energy, some protons are captured by the ^{13}C nucleus which then decays by gamma emission, giving off gamma rays with energies up to 11 MeV. A measurement of the angular distribution of the gamma rays emitted in this reaction is useful in describing the characteristics of energy levels in the residual nucleus ^{14}N.

Table 8-2 gives data for the number of 11 MeV gamma rays detected at each of six angles from 0° to 135°. The uncertainties are assumed to be purely statistical. These data are plotted in Figure 8-2 as a function of angle θ. There is a definite symmetry around $\theta = 90°$, and theoretical considerations of the reaction process predict that the data should be fitted well with a fourth-order Legendre polynomial with only even terms.

$$C = a_0 P_0(x) + a_2 P_2(x) + a_4 P_4(x) \qquad x = \cos \theta$$

Let us develop the method of least squares for an arbitrary function of order n

$$y(x) = a_0 X_0(x) + a_1 X_1(x) + \cdots a_n X_n(x) \qquad (8\text{-}19)$$

and consider the Legendre polynomial of Equation (8-16) as an illustration of the techniques to be used. The functions $X_j(x)$ may

Table 8-2 Experimental data for a determination of the angular distribution of gamma rays emitted from the reaction $^{13}C(p,\gamma)^{14}N$. The incident energy of the protons is $E_p = 4.5$ MeV, and the energy of the emitted capture gamma rays is $E_\gamma = 11$ MeV

Angle θ, degrees	Counts	Calculated fit
0	1352	1357 ± 24
45	927	893 ± 20
60	804	829 ± 19
90	889	879 ± 20
105	855	855 ± 20
135	881	893 ± 20

Coefficients for least-squares fit to a Legendre polynomial

a_0	a_1	a_2	a_3	a_4	χ_ν^2
908.	34.	195.	63.	158.	1.759
$\pm 12.$	$\pm 29.$	$\pm 39.$	$\pm 64.$	$\pm 49.$	
918.		233.		206.	0.788
$\pm 8.$		$\pm 18.$		$\pm 18.$	

Normalized Legendre polynomial:

1. $+ (0.254 \pm .020)P_2 + (0.224 \pm .014)P_4$

Errors quoted are probable errors, not standard deviations.

be terms from the power-series expansion of Equation (8-1) or from the Legendre polynomial of Equation (8-16).

Method of least squares With the function of Equation (8-19) expressed as a sum over n terms,

$$y(x) = \sum_{j=0}^{n} [a_j X_j(x)]$$

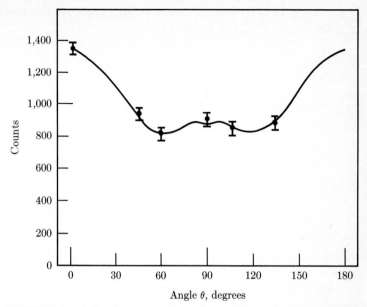

FIGURE 8-2 Graph of number of capture gamma rays from $^{13}C(p,\gamma)^{14}N$ detected vs. angle θ of detector with respect to incident proton beam direction. Calculated fit is for Legendre polynomial $918 + 233P_2(\cos\theta) + 206P_4(\cos\theta)$.

the definition of χ^2 becomes

$$\chi^2 \equiv \sum \left\{ \frac{1}{\sigma_i^2} [y_i - y(x_i)]^2 \right\} \tag{8-20}$$

We can express the determining equations for the method of least squares, which minimize χ^2, as a set of $n + 1$ equations (omitting σ_i^2 for the moment for clarity)

$$\frac{\partial}{\partial a_k} \Sigma[y_i - y(x_i)]^2 = \frac{\partial}{\partial a_k} \sum \left\{ y_i - \sum_{j=0}^{n} [a_j X_j(x_i)] \right\}^2 = 0$$

where the sum without indices runs from $i = 1$ to N; the sum over j runs from $j = 0$ to n; and k takes on all values from 0 to n.

The solution of these equations yields the values for the coefficients a_j for which χ^2 is a minimum.

As before, this yields a set of $n + 1$ simultaneous equations

$$\Sigma y_i X_0(x_i) = a_0 \Sigma X_0{}^2(x_i) + a_1 \Sigma X_1(x_i) X_0(x_i) \\ + \cdots + a_n \Sigma X_n(x_i) X_0(x_i)$$

$$\Sigma y_i X_1(x_i) = a_0 \Sigma X_0(x_i) X_1(x_i) + a_1 \Sigma X_1{}^2(x_i) \\ + \cdots + a_n \Sigma X_n(x_i) X_1(x_i)$$

. . . .

$$\Sigma y_i X_n(x_i) = a_0 \Sigma X_0(x_i) X_n(x_i) + a_1 \Sigma X_1(x_i) X_n(x_i) \\ + \cdots + a_n \Sigma X_n{}^2(x_i)$$

which can be expressed as (reinserting factors of $\sigma_i{}^2$)

$$\sum \left[\frac{1}{\sigma_i{}^2} y_i X_k(x_i) \right] = \sum_{j=0}^{n} \left\{ a_j \sum \left[\frac{1}{\sigma_i{}^2} X_j(x_i) X_k(x_i) \right] \right\} \qquad (8\text{-}21)$$

for all $k = 0, n$.

Matrix inversion The method of determinants which was introduced in Chapter 6 and discussed in Section 8-1 becomes cumbersome when used for polynomials of order 3 or more. A more general method for solving Equations (8-21) results from considering them as matrix equations which can be solved by rearrangement.

Equations (8-21) can be expressed in matrix form as the equivalence between a row matrix β and the product of the coefficient matrix a and a symmetric matrix α, all of degree $n + 1$

$$\beta_k = \sum_{j=0}^{n} (a_j \alpha_{jk}) \qquad k = 0, n$$
$$\beta = a\alpha \qquad\qquad\qquad\qquad (8\text{-}22)$$

where β and α are given in Equation (8-21).

$$\beta_k = \sum \left[\frac{1}{\sigma_i{}^2} y_i X_k(x_i) \right]$$
$$\alpha_{jk} = \sum \left[\frac{1}{\sigma_i{}^2} X_j(x_i) X_k(x_i) \right] \qquad (8\text{-}23)$$

The symmetric matrix $\boldsymbol{\alpha}$ is called the *curvature matrix* because of its relationship to the curvature of χ^2 in coefficient space. Combining Equations (8-19) and (8-20), the derivative of χ^2 with respect to any arbitrary coefficient a_j is given by

$$\frac{\partial \chi^2}{\partial a_j} = -2 \sum \left\{ \frac{1}{\sigma_i^2} [y_i - y(x_i)] X_j(x_i) \right\}$$

and the second cross partial derivative with respect to two such coefficients is

$$\frac{1}{2} \frac{\partial^2 \chi^2}{\partial a_j \, \partial a_k} = \sum \left[\frac{1}{\sigma_i^2} X_j(x_i) X_k(x_i) \right] = \alpha_{jk} \tag{8-24}$$

for all j and k.

If we multiply both sides of the matrix Equation (8-22) by another symmetrix matrix $\boldsymbol{\epsilon}$, which is equal to the inverse of the curvature matrix $\boldsymbol{\epsilon} = \boldsymbol{\alpha}^{-1}$, the result gives a solution for the coefficient matrix a directly since $\boldsymbol{\alpha}\boldsymbol{\alpha}^{-1} = 1$.

$$a = \beta\boldsymbol{\alpha}^{-1} = \beta\boldsymbol{\epsilon} \tag{8-25}$$

A more complete discussion of the use of matrices is given in Appendix B.

Equation (8-25) can be expressed in more conventional notation as

$$a_j = \sum_{k=0}^{n} (\epsilon_{jk}\beta_k) = \sum_{k=0}^{n} \left\{ \epsilon_{jk} \sum \left[\frac{1}{\sigma_i^2} y_i X_k(x_i) \right] \right\} \tag{8-26}$$

where the β_k are given by Equation (8-23). Note that since the curvature matrix $\boldsymbol{\alpha}$ is diagonally symmetric, that is, $\alpha_{jk} = \alpha_{kj}$, the inverse matrix $\boldsymbol{\epsilon}$ is equally symmetric: $\epsilon_{jk} = \epsilon_{kj}$. An analytical expression for the elements of the inverse matrix cannot be expressed simply. Instead, the elements of the curvature matrix $\boldsymbol{\alpha}$ must be assembled and the matrix inverted as discussed in Appendix B.

Estimation of errors　We can estimate the uncertainties in the coefficients a_j in the same way that we did in Section 6-5. The standard deviation σ_{a_j} for the uncertainty of determination of any coefficient a_j is the root sum square of the products of the standard deviation of each data point σ_i multiplied by the effect that data point has on the determination of the coefficient a_j [see Equation (4-9)].

$$\sigma_{a_j}{}^2 = \sum \left[\sigma_i{}^2 \left(\frac{\partial a_j}{\partial y_i} \right)^2 \right] \tag{8-27}$$

From Equation (8-26), the derivatives in the right-hand term are given by

$$\left(\frac{\partial a_j}{\partial y_i} \right) = \sum_{k=0}^{n} \left[\epsilon_{jk} \frac{1}{\sigma_i{}^2} X_k(x_i) \right]$$

and the weighted sum of the squares of the derivatives can be reduced to

$$\sum \left[\sigma_i{}^2 \left(\frac{\partial a_j}{\partial y_i} \right)^2 \right] = \sum_{k=0}^{n} \sum_{m=0}^{n} \left\{ \epsilon_{jk} \epsilon_{jm} \sum \left[\frac{1}{\sigma_i{}^2} X_k(x_i) X_m(x_i) \right] \right\}$$

$$= \sum_{k=0}^{n} \sum_{m=0}^{n} (\epsilon_{jk} \epsilon_{jm} \alpha_{km}) = \epsilon_{jj}$$

The inverse matrix $\boldsymbol{\epsilon} = \boldsymbol{\alpha}^{-1}$ is called the *error matrix* because it contains most of the information needed to estimate the errors.

$$\sigma_{a_j}{}^2 = \epsilon_{jj} \tag{8-28}$$

If the uncertainties in the data points are not known, they can be approximated from the data as in Equation (6-19)

$$\sigma_i{}^2 = \sigma^2 \simeq s^2 = \frac{1}{N - n - 1} \sum \left\{ y_i - \sum_{j=0}^{n} [a_j X_j(x_i)] \right\}^2 \tag{8-29}$$

where s^2 is the sample variance for the fit and $\nu = N - n - 1$ is the number of degrees of freedom after fitting N data points with $n + 1$ parameters. Combining Equations (8-27) to (8-29), the uncertainty in the coefficient a_j is given by

$$\sigma_{a_j}{}^2 \simeq s^2 \epsilon_{jj}(\sigma_i = 1) \tag{8-30}$$

where $\epsilon_{jj}(\sigma_i = 1)$ is the error matrix evaluated with $\sigma_i = 1$.

Program 8-2 LEGFIT Least-squares fit to a Legendre polynomial.

```
C     SUBROUTINE LEGFIT
C
C
C     PURPOSE
C       MAKE A LEAST-SQUARES FIT TO DATA WITH A LEGENDRE POLYNOMIAL
C         Y = A(1) + A(2)*X + A(3)*(3X**2-1)/2 + . . .
C           = A(1) * (1. + B(2)*X + B(3)*(3X**2-1)/2 + . . . )
C             WHERE X = COS(THETA)
C
C     USAGE
C       CALL LEGFIT (THETA, Y, SIGMAY, NPTS, NORDER, NEVEN, MODE, FTEST,
C         YFIT, A, SIGMAA, B, SIGMAB, CHISQR)
C
C     DESCRIPTION OF PARAMETERS
C       THETA  - ARRAY OF ANGLES (IN DEGREES) OF THE DATA POINTS
C       Y      - ARRAY OF DATA POINTS FOR DEPENDENT VARIABLE
C       SIGMAY - ARRAY OF STANDARD DEVIATIONS FOR Y DATA POINTS
C       NPTS   - NUMBER OF PAIRS OF DATA POINTS
C       NORDER - HIGHEST ORDER OF POLYNOMIAL (NUMBER OF TERMS - 1)
C       NEVEN  - DETERMINES ODD OR EVEN CHARACTER OF POLYNOMIAL
C                +1 FITS ONLY TO EVEN TERMS
C                 0 FITS TO ALL TERMS
C                -1 FITS ONLY TO ODD TERMS (PLUS CONSTANT TERM)
C       MODE   - DETERMINES MODE OF WEIGHTING LEAST-SQUARES FIT
C                +1 (INSTRUMENTAL) WEIGHT(I) = 1./SIGMAY(I)**2
C                 0 (NO WEIGHTING)  WEIGHT(I) = 1.
C                -1 (STATISTICAL)   WEIGHT(I) = 1./Y(I)
C       FTEST  - ARRAY OF VALUES OF F(L) FOR AN F TEST
C       YFIT   - ARRAY OF CALCULATED VALUES OF Y
C       A      - ARRAY OF COEFFICIENTS OF POLYNOMIAL
C       SIGMAA - ARRAY OF STANDARD DEVIATIONS FOR COEFFICIENTS
C       B      - ARRAY OF NORMALIZED RELATIVE COEFFICIENTS
C       SIGMAB - ARRAY OF STANDARD DEVIATIONS FOR RELATIVE COEFFICIENTS
C       CHISQR - REDUCED CHI SQUARE FOR FIT
C
C     SUBROUTINES AND FUNCTION SUBPROGRAMS REQUIRED
C       MATINV (ARRAY, NTERMS, DET)
C           INVERTS A SYMMETRIC TWO-DIMENSIONAL MATRIX OF DEGREE NTERMS
C           AND CALCULATES ITS DETERMINANT
C
C     MODIFICATIONS FOR FORTRAN II
C       OMIT DOUBLE PRECISION SPECIFICATIONS
C       CHANGE DCOS TO COSF IN STATEMENT 31
C       CHANGE DSQRT TO SQRTF IN STATEMENTS 156 AND 165
C
C     COMMENTS
C       DIMENSION STATEMENT VALID FOR NPTS UP TO 100 AND NORDER UP TO 9
C
```

Program 8-2 The method of calculation is illustrated in the computer routine LEGFIT of Program 8-2. This is a Fortran subroutine to make a least-squares fit to data with a Legendre polynomial, including all terms up to a maximum order n or just even or odd terms up to that maximum order. The input variables are THETA, Y, SIGMAY, NPTS, NORDER, NEVEN, MODE, and FTEST, and the output variables are YFIT, A, SIGMAA, B, SIGMAB, and CHISQR.

Program 8-2   LEGFIT   *(continued)*

```
      SUBROUTINE LEGFIT (THETA, Y, SIGMAY, NPTS, NORDER, NEVEN, MODE,
     1 FTEST, YFIT, A, SIGMAA, B, SIGMAB, CHISQR)
      DOUBLE PRECISION COSINE, P, BETA, ALPHA, CHISQ
      DIMENSION THETA(1), Y(1), SIGMAY(1), FTEST(1), YFIT(1),
     1 A(1), SIGMAA(1), B(1), SIGMAB(1)
      DIMENSION WEIGHT(100), P(100,10), BETA(10), ALPHA(10,10)
C
C         ACCUMULATE WEIGHTS AND LEGENDRE POLYNOMIALS
C
   11 NTERMS = 1
      NCOEFF = 1
      JMAX = NORDER + 1
   20 DO 40 I=1, NPTS
   21 IF (MODE) 22, 27, 29
   22 IF (Y(I)) 25, 27, 23
   23 WEIGHT(I) = 1. / Y(I)
      GO TO 31
   25 WEIGHT(I) = 1. / (-Y(I))
      GO TO 31
   27 WEIGHT(I) = 1.
      GO TO 31
   29 WEIGHT(I) = 1. / SIGMAY(I)**2
   31 COSINE = DCOS(.01745329252 * THETA(I))
      P(I,1) = 1.
      P(I,2) = COSINE
      DO 36 L=2, NORDER
      FL = L
   36 P(I,L+1) = ((2.*FL-1.)*COSINE*P(I,L) - (FL-1.)*P(I,L-1)) / FL
   40 CONTINUE
C
C         ACCUMULATE MATRICES ALPHA AND BETA
C
   51 DO 54 J=1, NTERMS
      BETA(J) = 0.
      DO 54 K=1, NTERMS
   54 ALPHA(J,K) = 0.
   61 DO 66 I=1, NPTS
      DO 66 J=1, NTERMS
      BETA(J) = BETA(J) + P(I,J)*Y(I)*WEIGHT(I)
      DO 66 K=J, NTERMS
      ALPHA(J,K) = ALPHA(J,K) + P(I,J)*P(I,K)*WEIGHT(I)
   66 ALPHA(K,J) = ALPHA(J,K)
C
C         DELETE FIXED COEFFICIENTS
C
   70 IF (NEVEN) 71, 91, 81
   71 DO 76 J=3, NTERMS, 2
      BETA(J) = 0.
      DO 75 K=1, NTERMS
      ALPHA(J,K) = 0.
   75 ALPHA(K,J) = 0.
   76 ALPHA(J,J) = 1.
      GO TO 91
   81 DO 86 J=2, NTERMS, 2
      BETA(J) = 0.
      DO 85 K=1, NTERMS
      ALPHA(J,K) = 0.
   85 ALPHA(K,J) = 0.
   86 ALPHA(J,J) = 1.
```

The variable NPTS $= N$ is the number of pairs of data points (θ_i, y_i). The variable NORDER $= n$ is the maximum order polynomial to which the subroutine will fit the data. The variable NEVEN

```
C
C         INVERT CURVATURE MATRIX ALPHA
C
   91 DO 95 J=1, JMAX
      A(J) = 0.
      SIGMAA(J) = 0.
      B(J) = 0.
   95 SIGMAB(J) = 0.
      DO 97 I=1, NPTS
   97 YFIT(I) = 0.
  101 CALL MATINV (ALPHA, NTERMS, DET)
      IF (DET) 111, 103, 111
  103 CHISQR = 0.
      GO TO 170
C
C         CALCULATE COEFFICIENTS, FIT, AND CHI SQUARE
C
  111 DO 115 J=1, NTERMS
      DO 113 K=1, NTERMS
  113 A(J) = A(J) + BETA(K)*ALPHA(J,K)
      DO 115 I=1, NPTS
  115 YFIT(I) = YFIT(I) + A(J)*P(I,J)
  121 CHISQ = 0.
      DO 123 I=1, NPTS
  123 CHISQ = CHISQ + (Y(I) - YFIT(I))**2 * WEIGHT(I)
      FREE = NPTS - NCOEFF
      CHISQR = CHISQ / FREE
C
C         TEST FOR END OF FIT
C
  131 IF (NTERMS - JMAX) 132, 151, 151
  132 IF (NCOEFF - 2) 133, 134, 141
  133 IF (NEVEN) 137, 137, 135
  134 IF (NEVEN) 135, 137, 135
  135 NTERMS = NTERMS + 2
      GO TO 138
  137 NTERMS = NTERMS + 1
  138 NCOEFF = NCOEFF + 1
      CHISQ1 = CHISQ
      GO TO 51
  141 FVALUE = (CHISQ1-CHISQ) / CHISQR
      IF (FTEST(NTERMS) - FVALUE) 134, 143, 143
  143 IF (NEVEN) 144, 146, 144
  144 NTERMS = NTERMS - 2
      GO TO 147
  146 NTERMS = NTERMS - 1
  147 NCOEFF = NCOEFF - 1
      JMAX = NTERMS
  149 GO TO 51
C
C         CALCULATE REMAINDER OF OUTPUT
C
  151 IF (MODE) 152, 154, 152
  152 VARNCE = 1.
      GO TO 155
  154 VARNCE = CHISQR
  155 DO 156 J=1, NTERMS
  156 SIGMAA(J) = DSQRT(VARNCE*ALPHA(J,J))
  161 IF (A(1)) 162, 170, 162
  162 DO 166 J=2, NTERMS
      IF (A(J)) 164, 166, 164
  164 B(J) = A(J) / A(1)
  165 SIGMAB(J) = B(J) * DSQRT((SIGMAA(J)/A(J))**2 + (SIGMAA(1)/A(1))**2
     1 - 2.*VARNCE*ALPHA(J,1)/(A(J)*A(1)))
  166 CONTINUE
      B(1) = 1.
  170 RETURN
      END
```

determines whether the subroutine will fit to all terms (NEVEN = 0), only even terms (NEVEN = +1 or any positive integer), or only odd terms (NEVEN = −1 or any negative integer) of the series.

The angles θ_i at which the observations were made are assumed to be stored in the array THETA, and the observations y_i (or C_i for Example 8-2) are assumed to be stored in the array Y with the same ordering.

The variable MODE determines the method of weighting the fit. For no weighting (or equal uncertainties $\sigma_i = \sigma$ for all data points), MODE = 0. For instrumental uncertainties, MODE = +1 (or any positive integer), and the standard deviations σ_i for the observations y_i are assumed to be stored in the array SIGMAY with the same ordering as for the data points. For statistical fluctuations, MODE = −1 (or any negative integer), and the standard deviations are calculated from the observations $\sigma_i^2 = y_i$. Note that SIGMAY is neither used nor altered for 0 or negative values of MODE, and for these cases the name of any array may be used in the calling statement for SIGMAY.

The variable FTEST is an array of n values used for testing the significance of each term in the polynomial. We will defer a discussion of this test until Chapter 10, but for now it will suffice to say that this test will enable us to terminate the order of our polynomial fit if we find that succeeding terms are not significant.

Statements 20–40 contain the accumulation of all the terms needed for the fitting procedure. The weighting factors for the N data points are calculated in statements 21–29 according to the value of MODE. The terms in the Legendre polynomial series P(I,L+1) are calculated in statements 31–36 from the iteration formula of Equation (8-18) and the values for $P_0(x)$ and $P_1(x)$ of Equations (8-17). The index I represents the index i of the observation. The second index represents the term in the power series

$$\mathrm{P(I,L+1)} = P_L(x_I)$$

The variable NMAX is equal to the maximum number of terms in the series. The variable FL is the floating point equivalent of L.

Statements 61–66 accumulate the sums for the terms of the matrices β and α of Equations (8-23).

$$\text{BETA}(\text{J}) = \beta_j = \sum \left[\frac{y_i}{\sigma_i^2} P_j(\cos \theta_i) \right] \qquad j = J - 1$$

$$\text{ALPHA}(\text{J},\text{K}) = \alpha_{jk}$$

$$= \sum \left[\frac{1}{\sigma_i^2} P_j(\cos \theta_i) P_k(\cos \theta_i) \right] \qquad k = K - 1$$

If some coefficients $\text{A}(\text{M})$ are to be omitted from the fit ($\text{NEVEN} \neq 0$), their effect on the matrices β and α are deleted in statements 71–86.

$$\text{BETA}(\text{M}) = \text{ALPHA}(\text{M},\text{K}) = \text{ALPHA}(\text{K},\text{M}) = 0$$

$$\text{ALPHA}(\text{M},\text{M}) = 1$$

Statement 101 calls the subroutine MATINV which inverts the matrix $\text{ALPHA} = \alpha$ as indicated in Equation (8-25) and deposits it back into $\text{ALPHA} = \alpha^{-1} = \epsilon$.

The coefficients $\text{A}(\text{J}) = a_{J-1}$ are calculated in statements 111–113 according to Equation (8-26). Using these coefficients, the fitted values of $\text{YFIT}(\text{I}) = y(x_I)$ are calculated in statements 111–115 according to Equation (8-16). Similarly, the value of χ^2 is accumulated in statements 121–123. The value of $\text{CHISQR} = \chi_\nu^2$ returned to the main program is the reduced chi-square

$$\chi_\nu^2 = \frac{1}{N - n - 1} \chi^2 = \frac{\chi^2}{\nu}$$

where $\text{FREE} = \nu = N - n - 1$ is the number of degrees of freedom.

The standard deviations for the uncertainties in the coefficients $\text{SIGMAA}(\text{J}) = \sigma_{a_{J-1}}$ are calculated in statements 151–156. If the uncertainties in the data points are not known ($\text{MODE} = 0$), the experimental variance $\text{VARNCE} = s^2$ is calculated in statement 154 according to Equation (8-29), using the fact that χ^2 and s^2 differ by only a factor of ν for this case. (If this value of s^2 is used for σ_i^2 in the weighting, $\chi_\nu^2 = 1$.) If the weighting factors have already been incorporated into the accumulated sums ($\text{MODE} \neq 0$), this factor is omitted by setting $\text{VARNCE} = 1.$ in statement 152.

Thus, the calculation follows Equation (8-28) with the sample variance s^2 estimated from Equation (8-29), if necessary.

Statements 131–149 allow us to terminate the series whenever our test of FVALUE $= F$ is less than the corresponding input test value FTEST. This F test will be discussed in detail in Chapter 10. The subroutine actually calculates the fit first for one term of the polynomial (statement 11 sets NTERMS $= 1$) and then adds successively higher-order terms until either the maximum order specified NORDER is reached or the test of F is satisfied.

Sample calculation   For example, in our experiment of Example 8-2, the data were fit with a fourth-order Legendre polynomial with the coefficients given in Table 8-2. The value of $\chi_\nu^2 = 0.438$ is acceptable, but the estimated errors for the coefficients of the odd terms are larger than the values of the terms themselves. This indicates that the terms are probably not significant and that we could improve the fit by specifying only even terms in the polynomial. In fact, the test values of F had to be unreasonably small to allow this fit.

The coefficients for the fit to even terms only is also given in Table 8-2. These values of the coefficients are not significantly different, but the fit is better, as indicated by the smaller values for χ^2 and for the estimated errors for the coefficients.

Normalized polynomials   If we are more interested in the shape of the curve than in the magnitude, the polynomial of Equation (8-19) should be rewritten in such a way as to separate the relative magnitude of the coefficients from the absolute magnitude.

$$y(x) = a_0 \left\{ 1 + \sum_{j=1}^{n} [b_j X_j(x_i)] \right\} \tag{8-31}$$

Evaluation of the coefficients b_j of Equation (8-31) is straightforward,

$$b_j = \frac{a_j}{a_0}$$

but determination of the uncertainties is more difficult.

From Equation (4-11), the uncertainty in the ratio of the two quantities a_j and a_0 is given by

$$\frac{\sigma_{b_j}{}^2}{b_j{}^2} = \frac{\sigma_{a_j}{}^2}{a_j{}^2} + \frac{\sigma_{a_0}{}^2}{a_0{}^2} - 2\frac{\sigma_{a_j a_0}{}^2}{a_j a_0} \tag{8-32}$$

where the uncertainties in a_j and a_0 are given by Equation (8-28). The covariance $\sigma_{a_j a_0}{}^2$ follows from a derivation similar to that of Equations (8-27) and (8-28).

$$\sigma_{a_j a_0}{}^2 = \sum \left[\sigma_i{}^2 \left(\frac{\partial a_j}{\partial y_i} \right) \left(\frac{\partial a_0}{\partial y_i} \right) \right] = \epsilon_{j0} \tag{8-33}$$

These calculations are illustrated in statements 161–166 of LEGFIT in Program 8-2. The relative coefficients $B(L) = b_L$ are evaluated in statement 164, and the uncertainties

SIGMAB(L) $= \sigma_{b_L}$

are evaluated in statement 165 of the DO loop starting at statement 162. Typical results are given in Table 8-2 for the data of Example 8-2.

SUMMARY

Least-squares fit to a power-series polynomial:

$$y(x) = a + bx + cx^2 + dx^3 + \cdots + mx^n$$

$$\Delta = \begin{vmatrix} \sum \dfrac{1}{\sigma_i{}^2} & \sum \dfrac{x_i}{\sigma_i{}^2} & \sum \dfrac{x_i{}^2}{\sigma_i{}^2} & \cdots \\[2ex] \sum \dfrac{x_i}{\sigma_i{}^2} & \sum \dfrac{x_i{}^2}{\sigma_i{}^2} & \sum \dfrac{x_i{}^3}{\sigma_i{}^2} & \cdots \\[2ex] \sum \dfrac{x_i{}^2}{\sigma_i{}^2} & \sum \dfrac{x_i{}^3}{\sigma_i{}^2} & \sum \dfrac{x_i{}^4}{\sigma_i{}^2} & \cdots \\[2ex] \cdots & \cdots & \cdots \end{vmatrix}$$

$$a = \frac{1}{\Delta} \begin{vmatrix} \sum \dfrac{y_i}{\sigma_i{}^2} & \sum \dfrac{x_i}{\sigma_i{}^2} & \sum \dfrac{x_i{}^2}{\sigma_i{}^2} & \cdots \\[2ex] \sum \dfrac{x_i y_i}{\sigma_i{}^2} & \sum \dfrac{x_i{}^2}{\sigma_i{}^2} & \sum \dfrac{x_i{}^3}{\sigma_i{}^2} & \cdots \\[2ex] \sum \dfrac{x_i{}^2 y_i}{\sigma_i{}^2} & \sum \dfrac{x_i{}^3}{\sigma_i{}^2} & \sum \dfrac{x_i{}^4}{\sigma_i{}^2} & \cdots \\[2ex] \cdots & \cdots & \cdots \end{vmatrix}$$

For the jth coefficient, replace the jth column in Δ with the first column in a.

Sample variance s^2:

$$s^2 = \frac{1}{N - n - 1} \Sigma[y_i - y(x_i)]^2$$

Orthogonal polynomials:

$$y(x) = a + b(x - \beta) + c(x - \gamma_1)(x - \gamma_2) + \cdots$$
$$a = \bar{y} \qquad b = \frac{\Sigma[y_i(x_i - \beta)]}{\Sigma(x_i - \beta)^2}$$
$$c = \frac{\Sigma[y_i(x_i - \gamma_1)(x_i - \gamma_2)]}{\Sigma[(x_i - \gamma_1)(x_i - \gamma_2)]^2}$$

For equally spaced values of $x_{i+1} - x_i = \Delta$ and constant $\sigma_i = \sigma$,

$$\beta = \tfrac{1}{2}(x_1 + x_N) \qquad \gamma = \beta \pm \Delta \sqrt{\tfrac{1}{12}(N^2 - 1)}$$
$$\delta = \beta, \quad \beta \pm \Delta \sqrt{\tfrac{1}{20}(3N^2 - 7)}$$

Legendre polynomials:

$$y(x) = \sum_{L=0}^{n} [a_L P_L(x)]$$
$$P_0(x) = 1 \qquad P_1(x) = x$$
$$P_L(x) = \frac{1}{L}[(2L - 1)x P_{L-1}(x) - (L - 1)P_{L-2}(x)]$$

Least-squares fit to an arbitrary function:

$$y(x) = \sum_{j=0}^{n} [a_j X_j(x)]$$
$$a_j = \sum_{k=0}^{n} (\epsilon_{jk}\beta_k)$$
$$\beta_k = \sum \left[\frac{1}{\sigma_i^2} y_i X_k(x_i)\right]$$
$$\boldsymbol{\epsilon} = \boldsymbol{\alpha}^{-1} \quad \text{where } \alpha_{jk} = \sum \left[\frac{1}{\sigma_i^2} X_j(x_i)X_k(x_i)\right]$$

Uncertainties in coefficients σ_{a_j}:

$$\sigma_{a_j}^2 = \epsilon_{jj} \simeq s^2 \epsilon_{jj}(\sigma_i = 1)$$

For normalized polynomials:

$$y(x) = a_0 \left\{ 1 + \sum_{j=1}^{n} [b_j X_j(x)] \right\}$$

$$\frac{\sigma_{b_j}{}^2}{b_j{}^2} = \frac{\sigma_{a_j}{}^2}{a_j{}^2} + \frac{\sigma_{a_0}{}^2}{a_0{}^2} - 2\frac{\sigma_{a_j a_0}{}^2}{a_j a_0}$$

$$\sigma_{a_j a_0}{}^2 = \epsilon_{j0} = \alpha_{j0}{}^{-1}$$

EXERCISES

8-1 Compare the discrepancies Δ_i of Example 8-1 with the experimental uncertainty s. How much larger than s is the largest value of Δ_i? How probable is such a discrepancy?

8-2 Write down the determinant Δ of Equations (8-5) for a cubic fitting function $y = a + bx + cx^2 + dx^3$.

8-3 Fit the data of Example 8-2 with a fourth-degree power-series polynomial.

8-4 Derive expressions for γ_1 and γ_2 of Equations (8-14).

8-5 Derive an expression for $P_4(\cos \theta)$.

8-6 Show that

$$\sum_{k=0}^{n} \sum_{m=0}^{n} (\epsilon_{jk}\epsilon_{jm}\alpha_{km}) = \epsilon_{jj}$$

8-7 Evaluate the uncertainties in the coefficients of the normalized Legendre polynomial of Example 8-2 ignoring the last term in Equation (8-32). Compare them with the uncertainties given in Example 8-2. Is the last term of Equation (8-32) significant?

8-8 Rewrite the pertinent section of LEGFIT to fit data with a power-series polynomial.

CHAPTER NINE

MULTIPLE REGRESSION

9-1 MULTIPLE LINEAR REGRESSION

The techniques of least-squares fitting which we have developed fall under the general name of regression analysis. Since we have so far considered only cases for which the polynomial $y = f(x)$ is linear in the parameters, we have been considering only linear regression, or multiple linear regression for polynomials more complex than straight lines and for linear functions of several variables. In this chapter we will generalize the methods of multiple linear regression to optimize the techniques of solution and to include functions which are not polynomials.

The method of determinants introduced in Chapter 6 for solving simultaneous equations is inadequate to handle the method of least squares for fitting data with more than the simplest of polynomials. The matrix method introduced in Section 8-3 is more sophisticated, but there exists an even more elegant approach utilizing matrix manipulation which is appropriate for use in multiple linear-regression analysis. Its advantage over straightforward matrix methods lies in the fact that it requires the solution of one fewer simultaneous equation, and the matrix manipulation is consequently reduced.

Elimination of first coefficient   Consider a general function which is linear in the coefficients

$$y = a_0 + a_1 X_1 + a_2 X_2 + \cdots + a_n X_n \qquad (9\text{-}1)$$

where the X_j are functions of independent variables. This is the expression of Equation (8-19) discussed in Section 8-3, with the first term modified to indicate the fact that it is a constant. The method of least squares results in a set of $n + 1$ simultaneous equations similar to those of Equations (8-21) (omitting σ_i^2 for the moment).

$$\begin{aligned}
\Sigma y_i &= a_0 N + a_1 \Sigma X_1 + a_2 \Sigma X_2 + \cdots + a_n \Sigma X_n \\
\Sigma y_i X_j &= a_0 \Sigma X_j + a_1 \Sigma X_1 X_j + a_2 \Sigma X_2 X_j \\
&\qquad\qquad + \cdots + a_n \Sigma X_n X_j
\end{aligned} \qquad (9\text{-}2)$$

The sums are taken over the N observations, considering the X_j to be functions $X_j(x_i)$.

We can simplify these equations and increase their symmetry if we consider the first equation to be a separable solution for the first term a_0 in the fitting function of Equation (9-1). That is, we can rewrite the first equation of Equations (9-2) in such a way that the coefficient a_0 is expressed explicitly as a function of all the other coefficients

$$\begin{aligned}
a_0 &= \frac{1}{N} \left[\Sigma y_i - a_1 \Sigma X_1 - a_2 \Sigma X_2 - \cdots - a_n \Sigma X_n \right] \\
&= \bar{y} - a_1 \bar{X}_1 - a_2 \bar{X}_2 - \cdots - a_n \bar{X}_n
\end{aligned} \qquad (9\text{-}3)$$

where $\bar{y}$ and the $\bar{X}_j$ are averages of y_i and the functions $X_j(x_i)$ over the N observations. After solving the remaining n simultaneous equations, we can then substitute the fitted values of the n coefficients a_j into Equation (9-3) to solve for the constant coefficient a_0.

Method of least squares   With this definition, the formula for χ^2 becomes

$$\chi^2 = \frac{1}{\sigma^2} \Sigma[(y_i - \bar{y}) - a_1(X_1 - \bar{X}_1)$$
$$- a_2(X_2 - \bar{X}_2) - \cdots - a_n(X_n - \bar{X}_n)]^2 \quad (9\text{-}4)$$

As before, the values of the coefficients for which χ^2 is a minimum are determined by setting the derivatives of χ^2 with respect to each of the n coefficients a_j equal to 0. This yields n simultaneous equations similar to Equations (9-2).

$$\Sigma(y_i - \bar{y})(X_k - \bar{X}_k)$$
$$= \sum_{j=1}^{n} [a_j \Sigma(X_j - \bar{X}_j)(X_k - \bar{X}_k)] \qquad k = 1, n \quad (9\text{-}5)$$

These equations can be simplified further by substituting the definitions for the linear-correlation coefficients r_{jk} given in Equations (7-10) to (7-15).

$$r_{jk} = \frac{s_{jk}{}^2}{s_j s_k} \qquad r_{jy} = \frac{s_{jy}{}^2}{s_j s_y}$$

$$s_{jk}{}^2 = \frac{1}{N-1} \Sigma[(X_j - \bar{X}_j)(X_k - \bar{X}_k)]$$

$$\qquad\qquad\qquad\qquad\qquad\qquad\qquad (9\text{-}6)$$

$$s_{jy}{}^2 = \frac{1}{N-1} \Sigma[(X_j - \bar{X}_j)(y_i - \bar{y})]$$

$$s_j{}^2 = \frac{1}{N-1} \Sigma(X_j - \bar{X}_j)^2 \qquad s_y{}^2 = \frac{1}{N-1} \Sigma(y_i - \bar{y})^2$$

Equations (9-5) become n simultaneous equations in b_j

$$r_{1y} = b_1 r_{11} + b_2 r_{21} + \cdots + b_n r_{n1}$$
$$r_{2y} = b_1 r_{12} + b_2 r_{22} + \cdots + b_n r_{n2}$$
$$\cdot \cdot \cdot \cdot \cdot \cdot \cdot \cdot \cdot \cdot \cdot \cdot \cdot \cdot \cdot \cdot \cdot \cdot \cdot \cdot \cdot \cdot \quad (9\text{-}7)$$
$$r_{ny} = b_1 r_{1n} + b_2 r_{2n} + \cdots + b_n r_{nn}$$

where the coefficients b_j are related to the coefficients a_j of the function of Equation (9-1) by

$$b_j = a_j \frac{s_j}{s_y} \quad (9\text{-}8)$$

and the diagonal terms in r are unity: $r_{jj} = 1$.

Matrix inversion As in Section 8-3, we consider Equations (9-7) to be a matrix equation (see Appendix B)

$$r_{ky} = \sum_{j=1}^{n} (b_j r_{jk}) \qquad k = 1, n$$
$$[r_y] = b\mathbf{r}$$

for which the solution is

$$b = [r_y]\mathbf{r}^{-1}$$
$$b_j = \sum_{k=1}^{n} (r_{ky} r_{jk}^{-1})$$

where $r_{jk}^{-1} = r_{kj}^{-1}$ is the jkth term of the matrix $\mathbf{r}^{-1}$, which is the inverse of the matrix $\mathbf{r}$ of the linear-correlation coefficients.

The *partial-regression coefficient* a_j of the original function of Equation (9-1) can be determined by substituting the definition of Equation (9-8).

$$a_j = \frac{s_y}{s_j} \sum_{k=1}^{n} (r_{ky} r_{jk}^{-1}) \quad (9\text{-}9)$$

The form of this equation is very similar to that of Equation

(8-26) except that the matrix components inside the summation are evaluated with respect to the centroid of function space.

Weighted fit If the uncertainties of the data points are not equal $\sigma_i \neq \sigma$, we must include the weighting factor σ_i^2 in the preceding derivation. Equations (9-2) to (9-5) will have factors of $1/\sigma_i^2$ in each term of their sums (see Section 10-1). The simultaneous equations of Equations (9-7) retain their form except that the definitions of Equations (9-6) for the correlation coefficients must be modified as in Equations (7-17).

$$r_{jk} = \frac{s_{jk}^2}{s_j s_k} \qquad r_{jy} = \frac{s_{jy}^2}{s_j s_y}$$

$$s_{jk}^2 = \frac{\dfrac{1}{N-1} \sum \left[\dfrac{1}{\sigma_i^2} (X_j - \bar{X}_j)(X_k - \bar{X}_k) \right]}{\dfrac{1}{N} \sum \dfrac{1}{\sigma_i^2}}$$

$$s_{jy}^2 = \frac{\dfrac{1}{N-1} \sum \left[\dfrac{1}{\sigma_i^2} (X_j - \bar{X}_j)(y_i - \bar{y}) \right]}{\dfrac{1}{N} \sum \dfrac{1}{\sigma_i^2}} \qquad (9\text{-}10)$$

$$s_j^2 = \frac{\dfrac{1}{N-1} \sum \left[\dfrac{1}{\sigma_i^2} (X_j - \bar{X}_j)^2 \right]}{\dfrac{1}{N} \sum \dfrac{1}{\sigma_i^2}}$$

$$s_y^2 = \frac{\dfrac{1}{N-1} \sum \left[\dfrac{1}{\sigma_i^2} (y_i - \bar{y})^2 \right]}{\dfrac{1}{N} \sum \dfrac{1}{\sigma_i^2}}$$

Similarly, the coefficients b_j in Equations (9-7) are related to the partial-regression coefficients a_j of Equation (9-1) as given in Equation (9-8) but with the new definitions for the standard deviations s_j and s_y. Since the variances and covariances are defined in terms of normalized weighting factors, this manner of weighting does not change their magnitudes drastically.

The matrix inversion leading to Equation (9-9) yields the same solution for the partial-regression coefficients.

$$a_j = \frac{s_y}{s_j} \sum_{k=1}^{n} (r_{ky} r_{jy}{}^{-1})$$

The constant term a_0 in Equation (9-1) is determined from the other coefficients, using weighted sums in Equation (9-3).

$$a_0 = \frac{\sum \left(\frac{1}{\sigma_i{}^2} \left\{ y_i - \sum_{j=1}^{n} [a_j X_j(x_i)] \right\} \right)}{\sum \frac{1}{\sigma_i{}^2}} \tag{9-11}$$

Estimation of errors   As in Sections 6-5 and 8-3, we can estimate the uncertainties in the coefficients a_j by combining the uncertainties of the individual data points σ_i with the effect which each data point has on the determination of the coefficients $\partial a_j / \partial y_i$.

$$\sigma_{a_j}{}^2 = \sum \left[\sigma_i{}^2 \left(\frac{\partial a_j}{\partial y_i} \right)^2 \right] \tag{9-12}$$

We can rewrite Equation (9-9) to facilitate evaluating the derivative in Equation (9-12).

$$a_j = \frac{1}{s_j} \sum_{k=1}^{n} \left(\frac{s_{ky}{}^2}{s_k} r_{jk}{}^{-1} \right)$$
$$\frac{\partial a_j}{\partial y_i} = \frac{1}{s_j} \sum_{k=1}^{n} \left[\frac{1}{s_k} r_{jk}{}^{-1} \left(\frac{\partial s_{ky}{}^2}{\partial y_i} \right) \right] \tag{9-13}$$

From the definitions of Equations (9-10), we can evaluate the derivative in the right-hand side of Equation (9-13).

$$\left(\frac{\partial s_{ky}}{\partial y_i} \right) = \frac{\frac{1}{N-1} \frac{1}{\sigma_i{}^2} (X_k - \bar{X}_k)}{\frac{1}{N} \sum \frac{1}{\sigma_i{}^2}} \tag{9-14}$$

The uncertainty in the coefficient a_j is obtained by combining Equations (9-12) to (9-14).

$$\sigma_{a_j}^2 = \frac{\dfrac{1}{(N-1)^2}\dfrac{1}{s_j^2}\sum\left(\sigma_i^2\left\{\sum_{k=1}^n\left[\dfrac{1}{s_k}r_{jk}^{-1}\dfrac{1}{\sigma_i^2}(X_k-\bar{X}_k)\right]\right\}^2\right)}{\left(\dfrac{1}{N}\sum\dfrac{1}{\sigma_i^2}\right)^2}$$

$$= \frac{\dfrac{1}{(N-1)^2}\dfrac{1}{s_j^2}\displaystyle\sum_{k=1}^n\sum_{m=1}^n\left\{\dfrac{1}{s_k s_m}r_{jk}^{-1}r_{jm}^{-1}\sum\left[\dfrac{1}{\sigma_i^2}(X_k-\bar{X}_k)(X_m-\bar{X}_m)\right]\right\}}{\left(\dfrac{1}{N}\sum\dfrac{1}{\sigma_i^2}\right)^2}$$

$$= \frac{\dfrac{1}{N-1}\dfrac{1}{s_j^2}\displaystyle\sum_{k=1}^n\sum_{m=1}^n(r_{jk}^{-1}r_{jm}^{-1}r_{km})}{\dfrac{1}{N}\sum\dfrac{1}{\sigma_i^2}}$$

By matrix multiplication (see Appendix B), this reduces to

$$\sigma_{a_j}^2 = \frac{\dfrac{1}{N-1}\dfrac{1}{s_j^2}r_{jj}^{-1}}{\dfrac{1}{N}\sum\dfrac{1}{\sigma_i^2}} \tag{9-15}$$

If the uncertainties of the data points are all equal, this can be simplified still further.

$$\sigma_{a_j}^2 = \frac{1}{N-1}\frac{1}{s_j^2}r_{jj}^{-1}\sigma^2 \simeq \frac{1}{N-1}\frac{s^2}{s_j^2}r_{jj}^{-1} \tag{9-16}$$

where

$$s^2 = \frac{1}{N-n-1}\Sigma[y_i - a_0 - \Sigma a_j X_j(x_i)]^2$$

$$= \frac{1}{N-n-1}\Sigma[(y_i-\bar{y}) - \Sigma a_j(X_j-\bar{X}_j)]^2 \tag{9-17}$$

By a similar development, the uncertainty σ_{a_0} in a_0 is

$$\sigma_{a_0}{}^2 = \frac{\dfrac{1}{N} + \dfrac{1}{N-1} \sum_{j=1}^{n} \sum_{k=1}^{n} \left(\bar{X}_j \bar{X}_k \frac{1}{s_j} \frac{1}{s_k} r_{jk}{}^{-1} \right)}{\dfrac{1}{N} \sum \dfrac{1}{\sigma_i{}^2}}$$

The numerator of Equation (9-15) is a function of the dispersion of the observations of the independent variables and is not a function of the observed data points y_i. This expression is analogous to the component ϵ_{jj} of the error matrix of Equation (8-30). The inverse linear-correlation matrix $\mathbf{r}^{-1}$ may be called the *reduced error matrix* in analogy with the error matrix $\boldsymbol{\epsilon} = \boldsymbol{\alpha}^{-1}$.

The denominator of Equation (9-15) is the reciprocal of the average variance $\overline{\sigma_i{}^2}$ of the data points y_i. Thus, the uncertainty σ_{a_j} in the determination of any coefficient a_j is the product of two terms; one represents the magnitude of the uncertainties of the data points, the other is the corresponding component of the reduced error matrix which determines the relative uncertainty in each coefficient.

Program 9-1 The method of calculation is illustrated in the computer routine REGRES of Program 9-1. This is a Fortran subroutine to make a multiple-regression (least-squares) fit to data with a function which is linear in the coefficients. The input variables are X, Y, SIGMAY, NPTS, NTERMS, M, and MODE, and the output variables are YFIT, A0, A, SIGMA0, SIGMAA, R, RMUL, CHISQR, and FTEST.

The values of the independent variable x_i are assumed to be stored in the array X, and the data for the dependent variable y_i are assumed to be stored in the array Y with identical ordering. The variable NPTS $= N$ represents the number of pairs of data points (x_i, y_i), and the variable NTERMS $= n$ represents the number of functional terms to be included in the function of Equation (9-1).

The variable MODE determines the method of weighting the fit as for Programs 8-1 and 8-2. For no weighting (or equal uncer-

Program 9-1   REGRES   Multiple linear-regression fit.

```
C      SUBROUTINE REGRES
C
C
C      PURPOSE
C        MAKE A MULTIPLE LINEAR REGRESSION FIT TO DATA WITH A SPECIFIED
C        FUNCTION WHICH IS LINEAR IN COEFFICIENTS
C
C      USAGE
C        CALL REGRES (X, Y, SIGMAY, NPTS, NTERMS, M, MODE, YFIT,
C         A0, A, SIGMA0, SIGMAA, R, RMUL, CHISQR, FTEST)
C
C      DESCRIPTION OF PARAMETERS
C        X       - ARRAY OF DATA POINTS FOR INDEPENDENT VARIABLE
C        Y       - ARRAY OF DATA POINTS FOR DEPENDENT VARIABLE
C        SIGMAY  - ARRAY OF STANDARD DEVIATIONS FOR Y DATA POINTS
C        NPTS    - NUMBER OF PAIRS OF DATA POINTS
C        NTERMS  - NUMBER OF COEFFICIENTS
C        M       - ARRAY OF INCLUSION/REJECTION CRITERIA FOR FCTN
C        MODE    - DETERMINES METHOD OF WEIGHTING LEAST-SQUARES FIT
C                  +1 (INSTRUMENTAL) WEIGHT(I) = 1./SIGMAY(I)**2
C                   0 (NO WEIGHTING) WEIGHT(I) = 1.
C                  -1 (STATISTICAL)  WEIGHT(I) = 1./Y(I)
C        YFIT    - ARRAY OF CALCULATED VALUES OF Y
C        A0      - CONSTANT TERM
C        A       - ARRAY OF COEFFICIENTS
C        SIGMA0  - STANDARD DEVIATION OF A0
C        SIGMAA  - ARRAY OF STANDARD DEVIATIONS FOR COEFFICIENTS
C        R       - ARRAY OF LINEAR CORRELATION COEFFICIENTS
C        RMUL    - MULTIPLE LINEAR CORRELATION COEFFICIENT
C        CHISQR  - REDUCED CHI SQUARE FOR FIT
C        FTEST   - VALUE OF F FOR TEST OF FIT
C
C      SUBROUTINES AND FUNCTION SUBPROGRAMS REQUIRED
C        FCTN (X, I, J, M)
C           EVALUATES THE FUNCTION FOR THE JTH TERM AND ITH DATA POINT
C           USING THE ARRAY M TO SPECIFY TERMS IN FUNCTION
C        MATINV (ARRAY, NTERMS, DET)
C           INVERTS A SYMMETRIC TWO-DIMENSIONAL MATRIX OF DEGREE NTERMS
C           AND CALCULATES ITS DETERMINANT
C
C      MODIFICATIONS FOR FORTRAN II
C        OMIT DOUBLE PRECISION SPECIFICATIONS
C        CHANGE DSQRT TO SQRTF IN STATEMENTS 72 AND 74
C        ADD F SUFFIX TO SQRT IN STATEMENTS 136 AND 146
C
C      COMMENTS
C        DIMENSION STATEMENT VALID FOR NPTS UP TO 100 AND NTERMS UP TO 10
C
```

tainties $\sigma_i = \sigma$ for all data points), MODE = 0. For instrumental uncertainties, MODE = +1 (or any positive integer), and the standard deviations σ_i for the data y_i are assumed to be stored in the array SIGMAY with the same ordering as for the data points. For statistical fluctuations, MODE = −1 (or any negative integer), and the standard deviations are calculated from the observations $\sigma_i^2 = 1/y_i$. As before, SIGMAY is neither used nor altered unless MODE > 0.

The functional terms $X_j(x_i)$ appearing in Equation (9-1)

Program 9-1   REGRES   *(continued)*

```
      SUBROUTINE REGRES (X, Y, SIGMAY, NPTS, NTERMS, M, MODE, YFIT,
     1 A0, A, SIGMA0, SIGMAA, R, RMUL, CHISQR, FTEST)
      DOUBLE PRECISION ARRAY, SUM, YMEAN, SIGMA, CHISQ, XMEAN, SIGMAX
      DIMENSION X(1), Y(1), SIGMAY(1), M(1), YFIT(1), A(1), SIGMAA(1),
     1 R(1)
      DIMENSION WEIGHT(100), XMEAN(10), SIGMAX(10), ARRAY(10,10)
C
C         INITIALIZE SUMS AND ARRAYS
C
   11 SUM = 0.
      YMEAN = 0.
      SIGMA = 0.
      CHISQ = 0.
      RMUL = 0.
      DO 17 I=1, NPTS
   17 YFIT(I) = 0.
   21 DO 28 J=1, NTERMS
      XMEAN(J) = 0.
      SIGMAX(J) = 0.
      R(J) = 0.
      A(J) = 0.
      SIGMAA(J) = 0.
      DO 28 K=1, NTERMS
   28 ARRAY(J,K) = 0.
C
C         ACCUMULATE WEIGHTED SUMS
C
   30 DO 50 I=1, NPTS
   31 IF (MODE) 32, 37, 39
   32 IF (Y(I)) 35, 37, 33
   33 WEIGHT(I) = 1. / Y(I)
      GO TO 41
   35 WEIGHT(I) = 1. / (-Y(I))
      GO TO 41
   37 WEIGHT(I) = 1.
      GO TO 41
   39 WEIGHT(I) = 1. / SIGMAY(I)**2
   41 SUM = SUM + WEIGHT(I)
      YMEAN = YMEAN + WEIGHT(I)*Y(I)
      DO 44 J=1, NTERMS
   44 XMEAN(J) = XMEAN(J) + WEIGHT(I)*FCTN(X,I,J,M)
   50 CONTINUE
   51 YMEAN = YMEAN / SUM
      DO 53 J=1, NTERMS
   53 XMEAN(J) = XMEAN(J) / SUM
      FNPTS = NPTS
      WMEAN = SUM / FNPTS
      DO 57 I=1, NPTS
   57 WEIGHT(I) = WEIGHT(I) / WMEAN
```

must be supplied by an additional function subprogram
FCTN$(\text{X,I,J,M}) = M_J(x_I)$. This routine is supplied by the user to
provide a fit to any desired function linear in its coefficients. The
array $M_j = X_j(x_i)$ is provided for use with this routine and will be
described in Section 9-2.

Statements 11–28 contain initialization procedures to zero
constants and arrays which will be used to accumulate sums over
data points i and/or coefficients j. In statements 30–57 the

Program 9-1 REGRES *(continued)*

```
C
C           ACCUMULATE MATRICES R AND ARRAY
C
   61 DO 67 I=1, NPTS
      SIGMA = SIGMA + WEIGHT(I)*(Y(I) - YMEAN)**2
      DO 67 J=1, NTERMS
      SIGMAX(J) = SIGMAX(J) + WEIGHT(I)*(FCTN(X,I,J,M) - XMEAN(J))**2
      R(J) = R(J) + WEIGHT(I)*(FCTN(X,I,J,M)-XMEAN(J))*(Y(I)-YMEAN)
      DO 67 K=1, J
   67 ARRAY(J,K) = ARRAY(J,K) + WEIGHT(I)*(FCTN(X,I,J,M)-XMEAN(J))*
     1 (FCTN(X,I,K,M)-XMEAN(K))
   71 FREE1 = NPTS - 1
   72 SIGMA = DSQRT(SIGMA / FREE1)
      DO 78 J=1, NTERMS
   74 SIGMAX(J) = DSQRT(SIGMAX(J) / FREE1)
      R(J) = R(J) / (FREE1 * SIGMAX(J) * SIGMA)
      DO 78 K=1, J
      ARRAY(J,K) = ARRAY(J,K) / (FREE1 * SIGMAX(J) * SIGMAX(K))
   78 ARRAY(K,J) = ARRAY(J,K)
C
C           INVERT SYMMETRIC MATRIX
C
   81 CALL MATINV (ARRAY, NTERMS, DET)
      IF (DET) 101, 91, 101
   91 A0 = 0.
      SIGMA0 = 0.
      RMUL = 0.
      CHISQR = 0.
      FTEST = 0.
      GO TO 150
C
C           CALCULATE COEFFICIENTS, FIT, AND CHI SQUARE
C
  101 A0 = YMEAN
  102 DO 108 J=1, NTERMS
      DO 104 K=1, NTERMS
  104 A(J) = A(J) + R(K) * ARRAY(J,K)
  105 A(J) = A(J) * SIGMA/SIGMAX(J)
  106 A0 = A0 - A(J)*XMEAN(J)
  107 DO 108 I=1, NPTS
  108 YFIT(I) = YFIT(I) + A(J)*FCTN(X,I,J,M)
  111 DO 113 I=1, NPTS
      YFIT(I) = YFIT(I) + A0
  113 CHISQ = CHISQ + WEIGHT(I)*(Y(I) - YFIT(I))**2
      FREEN = NPTS - NTERMS - 1
  115 CHISQR = CHISQ*WMEAN/FREEN
```

weighting factors $\text{WEIGHT}(\text{I}) = w_i = (1/\sigma_i^2)/(1/N)\Sigma(1/\sigma_i^2)$ are evaluated for the various modes and are used in the accumulation of the weighted means $\text{XMEAN}(\text{J}) = \bar{X}_J$, $\text{YMEAN} = \bar{y}$, and $\text{WMEAN} = (1/N)\Sigma(1/\sigma_i^2)$.

The standard deviations for the data points $\text{SIGMA} = s_y$ and the functional terms $\text{SIGMAX}(\text{J}) = \sigma_J$, and the linear-correlation coefficients $\text{R}(\text{J}) = r_{Jy}$ and $\text{ARRAY}(\text{J},\text{K}) = r_{JK}$ are accumulated in statements 61–67 and evaluated in statements 71–78 according to Equations (9-10). $\text{FREE1} = N - 1$ is the appropriate number of degrees of freedom. The subroutine MATINV (see Appendix B) is

Program 9-1 REGRES (*continued*)

```
C
C        CALCULATE UNCERTAINTIES
C
  121 IF (MODE) 122, 124, 122
  122 VARNCE = 1./WMEAN
      GO TO 131
  124 VARNCE = CHISQR
  131 DO 133 J=1, NTERMS
  132 SIGMAA(J) = ARRAY(J,J) * VARNCE / (FREE1*SIGMAX(J)**2)
      SIGMAAG(J) = SQRT (SIGMAA(J))
  133 RMUL = RMUL + A(J) * R(J) * SIGMAX(J)/SIGMA
      FREEJ = NTERMS
  135 FTEST = (RMUL/FREEJ) / ((1.-RMUL)/FREEN)
  136 RMUL = SQRT (RMUL)
  141 SIGMA0 = VARNCE / FNPTS
      DO 145 J=1, NTERMS
      DO 145 K=1, NTERMS
  145 SIGMA0 = SIGMA0 + VARNCE*XMEAN(J)*XMEAN(K)*ARRAY(J,K) /
     1 (FREE1*SIGMAX(J)*SIGMAX(K))
  146 SIGMA0 = SQRT (SIGMA0)
  150 RETURN
      END
```

called in statement 81 to invert the symmetric matrix

$$\text{ARRAY} = [r_{jk}]$$

and deposit it back into the matrix ARRAY.

The coefficients $\text{A(J)} = a_J$ are determined according to Equation (9-9) in statements 102–105. Using these coefficients, the constant term $\text{A0} = a_0$ is evaluated according to Equation (9-3) in statements 101–106. The fitted curve $\text{YFIT(I)} = y(x_I)$ of Equation (9-1) is accumulated in statements 107–108.

The value of χ^2 is evaluated directly in statements 111–113

$$\chi^2 = \sum \left(\frac{1}{\sigma_i^2} \left\{ y_i - \sum_{j=1}^{n} [a_j X_j(x_i)] \right\}^2 \right)$$

rather than according to the equivalent definition of Equation (9-4). The value of χ^2 which is returned to the calling program is the reduced chi-square $\text{CHISQR} = \chi^2/\nu$ where

$$\text{FREEN} = \nu = N - n - 1$$

is the number of degrees of freedom after fitting N data points with a function with n coefficients and one constant term.

The uncertainties in the coefficients $\text{SIGMAA(J)} = \sigma_{a_j}$ are evaluated in statements 121–132. If the uncertainties of the data points are equal $\sigma_i = \sigma$, $\text{MODE} = 0$ and the σ_{a_j} are evaluated according to Equation (9-16) where σ^2 is estimated by VARNCE $= s^2$ in statement 124 according to Equation (9-17). Under these conditions, CHISQR $= \chi_\nu^2$ is meaningless (its value is $\chi_\nu^2 = 1$); the value returned to the calling program is really the variance of the data points s^2. If the uncertainties σ_i are not equal (MODE $\neq 0$), the σ_{a_j} are evaluated according to Equation (9-15) with VARNCE $= N/\Sigma(1/\sigma_i{}^2) = \overline{\sigma_i{}^2}$ set equal to the average variance of the data points in statement 122. The uncertainty in the constant term σ_{a_0} is evaluated in statements 141–146.

The multiple linear-correlation coefficient RMUL $= R$ is evaluated in statements 133 and 136 according to the definition of Equation (7-18) in Section 7-2. This coefficient does not include the constant term which cannot be correlated with variations in the data. The calculation of FTEST in statement 135 for use in the F test will be discussed in Chapter 10.

9-2 POLYNOMIALS

The matrix techniques for making multiple linear-regression fits discussed in the previous section may be used to fit data with polynomials instead of the techniques discussed in Chapter 8. Equation (9-1) becomes

$$y(x) = a_0 + a_1 x + a_2 x^2 + \cdots + a_n x^n \tag{9-18}$$

which is equivalent to the function of Equation (8-1). The functions X_j are powers of x.

$$X_j(x_i) = x_i{}^j \tag{9-19}$$

The experiment of Example 8-1, for example, could be analyzed with this method. For a second-degree polynomial, however, the method of determinants is perfectly adequate and probably faster and more accurate. Both methods, of course, must yield identical results since they are based on the same method of least squares. One of the advantages of the correlation matrix-inversion method is that the parameters needed to determine the fit and estimate the errors are interesting statistical parameters: variances of the data and independent functions,

Program 9-2 FCTN (Power series) Fitting function for REGRES.

```
C      FUNCTION FCTN (POWER SERIES)
C
C      PURPOSE
C        EVALUATE TERMS OF POLYNOMIAL FUNCTION FOR REGRES SUBROUTINE
C           FCTN(X,I,J,JTERMS) = X(I)**JTERMS(J)
C
C      USAGE
C        RESULT = FCTN (X, I, J, JTERMS)
C
C      DESCRIPTION OF PARAMETERS
C        X      - ARRAY OF DATA POINTS FOR INDEPENDENT VARIABLE
C        I      - INDEX OF DATA POINTS
C        J      - INDEX OF TERM IN POLYNOMIAL FUNCTION
C        JTERMS - ARRAY OF POWERS
C
C      SUBROUTINES AND FUNCTION SUBPROGRAMS REQUIRED
C        NONE
C
C      MODIFICATIONS FOR FORTRAN II
C        NONE
C
       FUNCTION FCTN (X, I, J, JTERMS)
       DIMENSION X(1), JTERMS(1)
     1 JEXP = JTERMS(J)
       FCTN = X(I)**JEXP
       RETURN
       END
```

linear-correlation coefficients, and the reduced-error matrix which describes the relative accuracy and cross correlation of the coefficients. A determination of the estimated errors of the coefficients σ_{a_j} is awkward in the determinant method because of notation.

Program 9-2   The use of the subroutine REGRES of Program 9-1 for fitting data with the power-series polynomial of Equation (9-18) is illustrated with the computer routine FCTN of Program 9-2. This is a Fortran function subprogram which is given the name FCTN to make it compatible with the calling statement in REGRES and which is designed to evaluate the functions X_j of Equation (9-19) for that subroutine.

The input variables are X, I, J, and JTERMS, but only two of these (I and J) are supplied by the subroutine REGRES. The other two must be supplied by the main program that calls REGRES and must be transmitted via the chain of calling statements. The variable I $= i$ represents the data point for which the function is to be evaluated, and the variable $J = j$ indicates which term X_j, of the n terms comprising the fitted curve $y(x_i)$, is to be evaluated.

The independent quantities x_i of the data are assumed to be stored in the array X. This array is not used or altered by REGRES. Under some circumstances it may even be ignored by FCTN when the function can be expressed completely in terms of I and J. For example, in Example 6-1, $x_i = 1$.

The array JTERMS is assumed to contain information pertaining either to the functional behavior of the separate terms or to the desirability of excluding particular terms from the function. In Section 10-2 we will develop a test for the goodness of fit which will enable us to accept or reject individual terms of the fitting function to optimize the description of the data. For the present, let us assume that the calling program has suitable criteria for including only selected terms in the fitting function and transmits its choices based on these criteria via the array JTERMS.

In Program 9-2, for example, the power to which each term is raised is not the same as the number of that term $X_j(x_i) \neq x_i{}^j$. Instead, the appropriate power is obtained from the array JTERMS.

$$X_J(x_i) = x_i{}^{J'} \qquad J' = \text{JTERMS}(J) \qquad (9\text{-}20)$$

Thus, we can fit the data with a polynomial of degree n' but with only n terms where $n \leq n'$, indicating that some of the terms in the power series may be omitted. The calling program uses its criteria to establish which terms in the power series to include, and deposits in the array JTERMS the powers J' to which the values of the data points x_i are to be raised, as stated in Equation (9-20).

Program 9-3 Similarly, in the computer routine FCTN of Program 9-3, the terms to be included in the fitting polynomial are determined by the array JTERMS but in a different manner. This routine is a Fortran function subprogram designated FCTN like Program 9-2 for compatibility with REGRES, which evaluates terms for fitting with a Legendre polynomial series function.

$$y(\theta) = a_0 + \sum_{L=1}^{n} a_L P_L(\cos \theta)$$

Program 9-3 **FCTN** (Legendre polynomial) Fitting function for REGRES.

```
C      FUNCTION FCTN (LEGENDRE POLYNOMIAL)
C
C      PURPOSE
C        EVALUATE TERMS OF LEGENDRE POLYNOMIAL FOR REGRES SUBROUTINE
C          FCTN(X,I,J,JTERMS) = P(L,THETA)
C          L DETERMINED FROM J AND JTERMS
C          THETA = X(I)
C
C      USAGE
C        RESULT = FCTN (X, I, J, JTERMS)
C
C      DESCRIPTION OF PARAMETERS
C        X      - ARRAY OF DATA POINTS FOR INDEPENDENT VARIABLE
C        I      - INDEX OF DATA POINTS
C        J      - INDEX OF TERM IN POLYNOMIAL FUNCTION
C        JTERMS - ARRAY DETERMINING INCLUSION OR REJECTION OF TERMS
C
C      SUBROUTINES AND FUNCTION SUBPROGRAMS REQUIRED
C        NONE
C
C      MODIFICATIONS FOR FORTRAN II
C        OMIT DOUBLE PRECISION SPECIFICATIONS
C        CHANGE DCOS TO COSF IN STATEMENT 21
C
       FUNCTION FCTN (X, I, J, JTERMS)
       DOUBLE PRECISION COSINE, P1, P2, P3
       DIMENSION X(1), JTERMS(1)
C
C          FIND VALUE OF L FOR JTH TERM
C
    11 L = 0
       K = 0
    13 L = L + 1
       K = K + JTERMS(L)
    15 IF (K-J) 13, 21, 21
C
C          ITERATE TO FIND VALUE OF P(L)
C
    21 COSINE = DCOS(0.01745329252 * X(I))
       P1 = 1.
       P2 = COSINE
       IF (L-1) 25, 41, 31
    25 P2 = 1.
       GO TO 41
    31 DO 35 K=2, L
       FL = K
       P3 = ((2.*FL-1.)*COSINE*P2 - (FL-1.)*P1) / FL
       P1 = P2
    35 P2 = P3
    41 FCTN = P2
       RETURN
       END
```

Statements 21–35 use the iterative formula of Equation (8-18) to evaluate the Lth term of a Legendre polynomial at the angle specified by THETA(I) $= \theta_i$ (in degrees).

The value of L corresponding to the Jth term in the fitting function is determined in statements 11–15 assuming the array JTERMS contains a series of 1s and 0s indicating which terms of a regular Legendre polynomial series are to be included in the fitting function. As the loop searches through the contents of JTERMS, the variable K is increased by 1 each time a valid term (JTERMS(L) = 1) is encountered but not when the term is to be excluded (JTERMS(L) = 0). The search continues with L increasing until the Jth valid term is reached (K = J).

The important similarity between these two function subprograms is that they both supply to the fitting routine REGRES functions $y(x_i)$ which are composed of sums of terms but not necessarily evenly ordered terms. The function FCTN might even contain a wide variety of different types of functional terms from which the main calling program could choose by appropriately modifying the contents of the array JTERMS. This provides the needed flexibility for optimizing the form of the fitting function as well as its coefficients.

9-3 NON - LINEAR FUNCTIONS

In all the procedures developed so far we have assumed that the fitting function was linear in the coefficients. By that we mean that the function can be expressed as a sum of separate terms as in Equation (9-1), each multiplied by one and only one coefficient. How can we fit data with a function which is not linear in the coefficients? The method of least squares does not yield a straightforward analytical solution as it does for linear functions. In Chapter 11 we will investigate methods of searching parameter space for the values of the coefficients which will minimize the goodness-of-fit criterion χ^2. For now, let us consider approximate solutions using linear-regression techniques.

Linear approximation Some functions can be transformed in such a way as to appear to be linear functions. For example, if we were to fit data to a function of the form

$$y = ab^x \qquad (9-21)$$

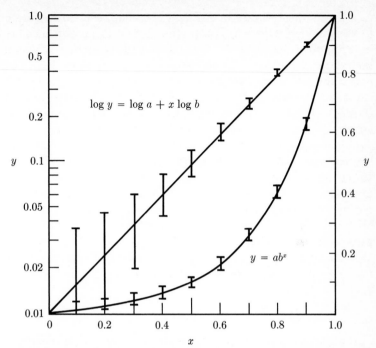

FIGURE 9-1 Graph of the function $y = ab^x$ for rectangular and logarithmic metric. Scale on the right is rectangular and refers to the curved line. Scale on the left is logarithmic and refers to the straight line. The function is evaluated for $a = 0.01$ and $b = 100$. The uncertainties shown are assumed to be equal in the rectangular metric $\sigma_i = 0.02$.

where a and b are constants, it seems reasonable to take logarithms (either natural logarithms ln or to the base 10 log) of both sides of the equation and fit the resulting straight line.

$$\log y = \log a + (\log b)x \tag{9-22}$$

The method of least squares minimizes the value of χ^2 with respect to each of the new coefficients $\log a$ and $\log b$ where χ^2 is given by

$$\chi^2 = \sum \left\{ \frac{1}{\sigma_i^2} \left[\log y_i - \log a - (\log b)x_i \right]^2 \right\} \tag{9-23}$$

If the uncertainties in the data are small, a fit to Equation (9-22) will yield a good fit to Equation (9-21). The problem comes in trying to optimize the fit for reasonable uncertainties. In effect, the reparameterization of Equation (9-22) overemphasizes the uncertainties for large values of y_i and tends to optimize the fit for the small values of y_i. This trend is illustrated in Figure 9-1 which shows the function of Equation (9-21) graphed both in a rectangular metric and on a logarithmic scale. The uncertainties of the data points are assumed to be equal as indicated on the rectangular scale, but they appear to be unequal on the logarithmic scale. In particular, the uncertainties are larger where the values of y_i are smaller. By fitting Equation (9-22) instead of Equation (9-21), we are assuming that the uncertainties are equal on the logarithmic scale and, therefore, we are underemphasizing the uncertainties for small values of y_i.

To compensate for this trend, we must modify the weighting factor σ_i appearing in the expression for χ^2.

$$\sigma_i' = \frac{d(\log y_i)}{dy_i}\sigma_i = \frac{\log e}{y_i}\sigma_i \tag{9-24}$$

If the data points have equal uncertainties $\sigma_i = \sigma$, these redefined uncertainties show just the variation illustrated in Figure 9-1. In general, if we apply the operation f to the data so that we fit the function $f(y)$ rather than y, the uncertainties σ_i must be equally compensated.

$$\sigma_i' = \frac{df(y_i)}{dy_i}\sigma_i \tag{9-25}$$

Similarly, if we were to fit to the function,

$$y = ae^{bx} \tag{9-26}$$

the fitting function would be

$$\ln y = \ln a + bx$$

and the uncertainties would be modified as in Equation (9-24).

Estimated errors If we modify the fitting function so that instead of fitting the data points y_i with coefficients a, b, . . . we fit modified data points $y'_i = f(y_i)$ with coefficients a', b', . . . , then our estimates of the errors in the coefficients will pertain to uncertainties in the modified coefficients a', b' rather than to the desired coefficients a, b, ... If the relationship between the two sets of coefficients is defined to be

$$a' = f_a(a) \qquad b' = f_b(b)$$

then the correspondence between the uncertainties σ'_a, σ'_b, . . . in the modified coefficients and the uncertainties σ_a, σ_b, . . . in the desired coefficients is obtained in a manner similar to that for σ'_i and σ_i in Equation (9-25).

$$\sigma'_a = \frac{df_a(a)}{da}\sigma_a \qquad \sigma'_b = \frac{df_b(b)}{db}\sigma_b \qquad (9\text{-}27)$$

Thus, if the modified coefficient is $a' = \ln a$, the estimated error in a is determined from the estimated error in a', according to Equation (9-27) with $f_a(a) = \ln (a)$.

$$\sigma'_a = \frac{d(\ln a)}{da}\sigma_a = \frac{\sigma_a}{a}$$

Values of χ^2 for testing the goodness of fit should be determined from the original uncertainties of the data σ_i and from the unmodified equation, although Equation (9-23) should give approximately equivalent results using the modified uncertainties σ'_i for weighting.

EXAMPLE 9-1 Consider the experiment of Example 6-2. The activity of a radioactive source is measured as a function of time and we wish to determine the initial activity and the rate of decay. In Chapter 6 these data were fit with a straight line, but theoretical considerations dictate that the function should be a decreasing exponential

$$C = C_0 e^{-t/\tau}$$

which is the same form as the function of Equation (9-26).

The weighting factor in the fourth column of Table 6-2 is $1/\sigma_i^2 \simeq 1/C_i$ which must be modified according to Equation (9-24).

$$w_i' = \frac{1}{\sigma_i'^2} = C^2 \frac{1}{\sigma_i^2} \simeq C_i \tag{9-28}$$

Table 9-1 illustrates the calculations necessary to fit the data

Table 9-1 Calculation of a fit to the data of Example 6-2 with an exponential function $C = C_0 e^{-t/\tau}$, approximated by fitting $y = a + bx$ where $y = \ln C$, $a = \ln C_0$, $b = -1/\tau$, $x = t$, and $w_i = C_i$

i	x_i	w_i	y_i	$w_i x_i$	$w_i x_i^2$	$w_i x_i y_i$	$w_i y_i$	$C_0 e^{-t/\tau}$
1	0	106	4.663	0	0	0	494	112.0
2	15	80	4.382	1200	18,000	5,260	351	97.9
3	30	98	4.585	2940	88,200	13,500	449	85.5
4	45	75	4.317	3380	151,900	14,500	324	74.7
5	60	74	4.304	4440	266,400	19,100	318	65.3
6	75	73	4.290	5480	410,600	23,500	313	57.0
7	90	49	3.892	4410	396,900	17,200	191	49.9
8	105	38	3.638	3990	419,000	14,500	138	43.6
9	120	37	3.611	4440	532,800	16,000	130	38.1
10	135	22	3.091	2970	401,000	9,250	68	33.3
SUM		652	40.773	33,250	2,685,000	132,800	2776	

$\Delta = \Sigma w_i x_i^2 \Sigma w_i - (\Sigma w_i x_i)^2 = 2,685,000(652) - (33,250)^2 = 6.46 \times 10^8$

$a = \dfrac{\Sigma w_i x_i^2 \Sigma w_i y_i - \Sigma w_i x_i \Sigma w_i x_i y_i}{\Delta} = \dfrac{2,685,000(2776) - 33,250(132,800)}{6.46 \times 10^8}$

$\quad = 4.72$

$b = \dfrac{\Sigma w_i \Sigma w_i x_i y_i - \Sigma w_i x_i \Sigma w_i y_i}{\Delta} = \dfrac{652(132,800) - 33,250(2776)}{6.46 \times 10^8} = -0.0090$

$\sigma_a^2 = \dfrac{\Sigma w_i x_i^2}{\Delta} = \dfrac{2.685 \times 10^6}{6.46 \times 10^8} = 0.00417 \qquad \sigma_a = 0.0645$

$\sigma_b^2 = \dfrac{\Sigma w_i}{\Delta} = \dfrac{652}{6.46 \times 10^8} = 0.99 \times 10^{-6} \qquad \sigma_b = 1.0 \times 10^{-3}$

$a = \ln C_0 \qquad C_0 = e^a = 112 \qquad \sigma_{C0} = C_0 \sigma_a = 7.2$

$b = -\dfrac{1}{\tau} \qquad \tau = -\dfrac{1}{b} = 111 \text{ sec} \qquad \sigma_\tau = \tau^2 \sigma_b = 12 \text{ sec}$

with a straight line of the form

$$\ln C_i = \ln C_0 - \frac{t}{\tau}$$

with weighting factors determined according to Equation (9-28). This is a case where the uncertainties were not equal even in the original metric (Poisson statistics), but the modification of the weighting is the same as above.

SUMMARY

Linear function:

$$y(x) = a_0 + \sum_{j=1}^{n} [a_j X_j(x)]$$

Covariance s_{jk}^2:

$$s_{jk}^2 = \frac{\dfrac{1}{N-1} \sum \left[\dfrac{1}{\sigma_i^2} (X_j - \bar{X}_j)(X_k - \bar{X}_k) \right]}{\dfrac{1}{N} \sum \dfrac{1}{\sigma_i^2}}$$

$$\bar{X}_j = \frac{\sum \left[\dfrac{1}{\sigma_i^2} X_j(x_i) \right]}{\sum \dfrac{1}{\sigma_i^2}} \qquad \bar{y} = \frac{\sum \left(\dfrac{1}{\sigma_i^2} y_i \right)}{\sum \dfrac{1}{\sigma_i^2}}$$

Variance:

$$s_j^2 = s_{jj}^2$$

Linear-correlation coefficient:

$$r_{jk} = \frac{s_{jk}^2}{s_j s_k}$$

Partial linear-regression coefficient a_j:

$$a_0 = \bar{y} - \sum_{j=1}^{n} (a_j \bar{X}_j)$$

$$a_j = \frac{s_y}{s_j} \sum_{k=1}^{n} (r_{ky} r_{jk}^{-1})$$

Uncertainty in coefficient σ_{a_j}:

$$\sigma_{a_j}{}^2 \simeq \frac{\dfrac{1}{N-1}\dfrac{1}{s_j{}^2}r_{jj}{}^{-1}}{\dfrac{1}{N}\sum\dfrac{1}{\sigma_i{}^2}}$$

$$\sigma_{a_0}{}^2 \simeq \frac{\dfrac{1}{N} + \dfrac{1}{N-1}\displaystyle\sum_{j=1}^{n}\left[\bar{X}_j{}^2 \dfrac{1}{s_j{}^2}r_{jj}{}^{-1} + \sum_{k=1}^{n}\left(\bar{X}_j\bar{X}_k \dfrac{1}{s_j s_k}r_{jk}{}^{-1}\right)\right]}{\dfrac{1}{N}\sum\dfrac{1}{\sigma_i{}^2}}$$

Non-linear function:

$$y' = f_y(y) = \sum_{j=1}^{n}[f_j(a_j)X_j'(x)]$$

$$\sigma_i' = \frac{df_y(y_i)}{dy_i}\sigma_i' \qquad \sigma_{a_j}' = \frac{df_j(a_j)}{da_j}\sigma_{a_j}$$

EXERCISES

9-1 Would you expect the methods of LEGFIT and REGRES to give identical results (within round-off error)?

9-2 Use REGRES to fit the data of Example 8-1.

9-3 Since the range of r_{jy} and r_{jk} is from -1 to $+1$, what can you say about the range of the coefficients b_j?

9-4 Use REGRES to fit the data of Example 8-2.

9-5 What transformation should be made to make a linear fit to data with a function $y = 1/(a + bx)$? What is the expression of Equation (9-25) for the modified uncertainties?

CHAPTER TEN

GOODNESS OF FIT

The method of least squares is built on the hypothesis that the optimum description of a set of data is one which minimizes the weighted sum of squares of deviations of the data y_i from the fitting function $y(x_i)$. This sum is characterized by the variance of the fit s^2, which is an estimate of the variance of the data σ^2. For a function with a constant term and n coefficients fit to N data points,

$$s^2 = \frac{\dfrac{1}{N - n - 1} \sum \left\{ \dfrac{1}{\sigma_i^2} \left[y_i - y(x_i) \right]^2 \right\}}{\dfrac{1}{N} \sum \dfrac{1}{\sigma_i^2}}$$

where the factor $\nu = N - n - 1$ is the number of degrees of freedom left after fitting N data points to the $n + 1$ parameters.

The weighting factor w_i for each data point is the inverse of the variance σ_i^2 which describes the uncertainty of that data point, normalized to the average of all the weighting factors.

$$w_i = \frac{1/\sigma_i^2}{(1/N)\Sigma(1/\sigma_i^2)} \tag{10-1}$$

The reason for including this weighting factor in all the calculations is that this formula yields the maximum likelihood that the fitting function represents the parent distribution, as derived in Section 6-2 leading to Equation (6-6).

The variance of the fit s^2 is also characterized by the statistic χ^2 defined in Equations (8-3) and (8-20) for polynomials.

$$\chi^2 \equiv \sum \left\{ \frac{1}{\sigma_i^2} [y_i - y(x_i)]^2 \right\} \tag{10-2}$$

The relationship between s^2 and χ^2 can be seen most easily by comparing s^2 with the reduced chi-square χ_ν^2

$$\chi_\nu^2 = \frac{\chi^2}{\nu} = \frac{s^2}{\overline{\sigma_i^2}}$$

where $\overline{\sigma_i^2}$ is the weighted average of the individual variances

$$\overline{\sigma_i^2} = \frac{\dfrac{1}{N} \sum \left(\dfrac{1}{\sigma_i^2} \sigma_i^2 \right)}{\dfrac{1}{N} \sum \dfrac{1}{\sigma_i^2}} = \frac{1}{\dfrac{1}{N} \sum \dfrac{1}{\sigma_i^2}} \tag{10-3}$$

and is equivalent to σ^2 if the uncertainties are equal $\sigma_i = \sigma$.

The parent variance of the data σ^2 is a characteristic of the dispersion of the data about the parent distribution and is not descriptive of the fit. The estimated variance of the fit s^2, however, is characteristic of both the spread of the data and the accuracy of the fit. The definition of χ^2 as the ratio of the estimated variance s^2 to the parent variance σ^2 (times the number of degrees of freedom ν) makes it a convenient measure of the goodness of fit.

If the fitting function is a good approximation to the parent function, then the estimated variance s^2 should agree well with the parent variance σ^2, and the value of the reduced chi-square should be approximately unity $\chi_\nu^2 = 1$. If the fitting function is not appropriate for describing the data, the deviations will be larger and the estimated variance will be too large, yielding a value of χ_ν^2 greater than 1. A value of χ_ν^2 less than 1 does not indicate an improvement of the fit, however; it is simply a consequence of the fact that there exists an uncertainty in the determination of s^2, and the observed values of χ_ν^2 will fluctuate from experiment to experiment.

Distribution of χ^2   The probability distribution function for χ^2 with ν degrees of freedom is given by

$$P_x(x^2, \nu) = \frac{(x^2)^{\frac{1}{2}(\nu-2)} e^{-x^2/2}}{2^{\nu/2} \Gamma(\nu/2)} \tag{10-4}$$

where the gamma function $\Gamma(z)$ has already been introduced in Program 7-2 as the factorial function for integral and half-integral arguments. The distribution of Equation (10-4) is called the chi-square distribution. It is derived in many texts on statistics[1] but we will simply quote the results here.

If the function of the parent population is denoted by $y_0(x)$, the value of χ_0^2 determined from the parameters of the parent function

$$\chi_0^2 = \sum \left\{ \frac{1}{\sigma_i^2} [y_i - y_0(x_i)]^2 \right\}$$

is distributed according to Equation (10-4) with $\nu = N$ degrees of freedom. If the function $y(x)$ used in the determination of χ^2 contains $n + 1$ parameters (n coefficients and one constant term), the resultant value of χ^2 from Equation (10-2) is distributed according to Equation (10-4) with $\nu = N - n - 1$ degrees of

[1] See Pugh and Winslow, sec. 12–5, for a derivation.

freedom. What is more important for our purposes is the integral probability $P_\chi(\chi^2,\nu)$ which is defined as the integral of the distribution function $P_x(x^2,\nu)$ of Equation (10-4) between $x^2 = \chi^2$ and $x^2 = \infty$.

$$P_\chi(\chi^2,\nu) = \int_{\chi^2}^{\infty} P_x(x^2) \, dx^2 \tag{10-5}$$

This function is tabulated and graphed in Appendix C-4 as a function of χ_ν^2. It describes the probability that a random set of N data points would yield a value of χ_0^2 as large or larger when compared with the parent function.

If the fitting function is a good approximation to the parent function, the experimental value of χ_ν^2 should be average and the probability of Equation (10-5) should be approximately 0.5. For larger values of χ^2, the probability of obtaining such a large value of χ^2 from the correct fitting function is smaller, indicating that the fitting function actually used may not be appropriate. Although there is an ambiguity in interpreting the probability, because even correct fitting functions can yield large values of χ^2 occasionally, in general the probability of Equation (10-5) is either reasonably close to 0.5, indicating a reasonable fit, or unreasonably small, indicating a bad fit.

For most purposes, the reduced chi-square χ_ν^2 is an adequate measure of the probability directly. The probability is reasonably close to 0.5 so long as χ_ν^2 is reasonably close to 1, that is, less than 1.5.

Program 10-1 The probability function $P_\chi(\chi^2,\nu)$ of Equation (10-5) can be computed by expanding the integral as in PCORRE of Program 7-1. For even values of ν, the integral can be expanded into a simple analytical expression.

$$P_\chi(\chi^2,\nu) = e^{-\chi^2/2} \sum_{i=0}^{I} \frac{(\chi^2/2)^i}{i!} \qquad I = \tfrac{1}{2}(\nu - 2), \; \nu \text{ even}$$

For odd values of ν, the integral cannot be evaluated analytically, but must be expanded as an infinite sum by expanding the exponential.

$$P_\chi(\chi^2,\nu) = 1 - \frac{1}{\Gamma(I+1)} \sum_{i=0}^{\infty} \left[(-1)^i \frac{(\chi^2/2)^{I+i+1}}{i!(I+i+1)} \right]$$

$$\nu \text{ odd}$$

The computation of $P_\chi(\chi^2,\nu)$ is illustrated in the computer routine PCHISQ of Program 10-1. The sum in the expansion is accumulated in statements 31–34 for ν even or in statements 51–56 for ν odd. The value of PCHISQ $= P_\chi(\chi^2,\nu)$ is evaluated in statement 35 or 57 and returned to the main program as the value of the function.

EXAMPLE 10-1 The subroutine POLFIT of Program 8-1, for example, contains a calculation of χ_ν^2 for evaluating the goodness of fit to the data of a polynomial.

$$y(x) = \sum_{j=0}^{n} a_j x^j$$

Equation (10-2) can be expanded for computation.

$$\chi^2 = \sum \left(\frac{1}{\sigma_i^2} y_i^2 \right) - 2 \sum_{j=0}^{n} \left[a_j \sum \left(\frac{1}{\sigma_i^2} y_i x^j \right) \right]$$
$$+ \sum_{j=0}^{n} \sum_{k=0}^{n} \left[a_j a_k \sum \left(\frac{1}{\sigma_i^2} x^{j+k} \right) \right]$$

The first term in this expansion is accumulated in statement 49 as part of the DO loop over the N data points extending from statements 21 to 50. The second and third terms are accumulated in statements 71–75, using the already accumulated sums over data points SUMY(J) $= \Sigma(y_i x^{J-1}/\sigma_i^2)$.

This method is convenient because it utilizes sums which have already been accumulated, but it is not a good method in general because it involves summing over many terms which tend to cancel and the residual has a diminished accuracy from computation.

Program 10-1 PCHISQ Integral probability function $P_\chi(\chi^2, \nu)$ of exceeding χ^2.

```
C      FUNCTION PCHISQ
C
C      PURPOSE
C        EVALUATE PROBABILITY FOR EXCEEDING CHI SQUARE
C
C      USAGE
C        RESULT = PCHISQ (CHISQR, NFREE)
C
C      DESCRIPTION OF PARAMETERS
C        CHISQR - COMPARISON VALUE OF REDUCED CHI SQUARE
C        NFREE  - NUMBER OF DEGREES OF FREEDOM
C
C      SUBROUTINES AND FUNCTION SUBPROGRAMS REQUIRED
C        GAMMA(X)
C          CALCULATES GAMMA FUNCTION FOR INTEGERS AND HALF-INTEGERS
C
C      MODIFICATIONS FOR FORTRAN II
C        OMIT DOUBLE PRECISION SPECIFICATIONS
C        CHANGE DEXP TO EXPF IN STATEMENT 35
C        CHANGE DABS TO ABSF IN STATEMENT 55
C
C      COMMENTS
C        CALCULATION IS APPROXIMATE FOR NFREE ODD AND
C          CHI SQUARE GREATER THAN 50
C
```

The value returned to the main program is $\chi_\nu^2 = \chi^2/\nu$ which is evaluated in statements 76–77. When applied to the data of Example 8-1, the result is $\chi_\nu^2 = 0.056$, which indicates that our estimate of the uncertainty σ in the data points is probably incorrect. In this case we must use the expression of Equation (8-7) to estimate s^2, and $\chi_\nu^2 = 1$ is meaningless.

EXAMPLE 10-2   Similarly, the subroutine REGRES of Program 9-1 contains a calculation of χ_ν^2 for evaluating the goodness of fit to the data of a general series.

$$y(x) = a_0 + \sum_{j=1}^{n} a_j X_j(x_i)$$

The value of χ_ν^2 is accumulated in statements 111–115 according to Equation (10-2) where YFIT(I) $= y(x_i)$; WEIGHT(I) $= w_i$ of Equation (10-1); WMEAN $= 1/\overline{\sigma_i^2}$ of Equation (10-3); and FREEN $= \nu = N - n - 1$ is the number of degrees of freedom.

Note that if the uncertainties of the data points σ_i are not

Program 10-1 PCHISQ *(continued)*

```
      FUNCTION PCHISQ (CHISQR, NFREE)
      DOUBLE PRECISION Z, TERM, SUM
   11 IF (NFREE) 12, 12, 14
   12 PCHISQ = 0.
      GO TO 60
   14 FREE = NFREE
      Z = CHISQR*FREE/2.
      NEVEN = 2*(NFREE/2)
      IF (NFREE - NEVEN) 21, 21, 41
C
C        NUMBER OF DEGREES OF FREEDOM IS EVEN
C
   21 IMAX = NFREE/2
      TERM = 1.
      SUM = 0.
   31 DO 34 I=1, IMAX
      FI = I
      SUM = SUM + TERM
   34 TERM = TERM * Z/FI
   35 PCHISQ = SUM * DEXP(-Z)
      GO TO 60
C
C        NUMBER OF DEGREES OF FREEDOM IS ODD
C
   41 IF (Z - 25) 44, 44, 42
   42 Z = CHISQR * (FREE-1.)/2.
      GO TO 21
   44 PWR = FREE/2.
      TERM = 1.
      SUM = TERM/PWR
   51 DO 56 I=1, 1000
      FI = I
      TERM = -TERM * Z/FI
      SUM = SUM + TERM/(PWR+FI)
   55 IF (DABS(TERM/SUM) - .00001) 57, 57, 56
   56 CONTINUE
   57 PCHISQ = 1. - (Z**PWR)*SUM/GAMMA(PWR)
   60 RETURN
      END
```

known or estimated from other experiments, we cannot evaluate χ_ν^2 by estimating the average uncertainty $\overline{\sigma_i}$ from the data because χ_ν^2 is the ratio of the estimated variance s^2 to the true variance $\overline{\sigma_i^2}$. Since our best estimate of $\overline{\sigma_i^2}$ is s^2, this would yield a value of exactly 1 for χ_ν^2. This is indicated in Program 9-1 in statements 121–124. If MODE = 0 ($\sigma_i = \sigma$, unknown), VARNCE is set equal to CHISQR for further calculation. Otherwise it is set equal to $1./\text{WMEAN} = \overline{\sigma_i^2}$.

Program 10-2 The calculation of χ^2 in Equation (10-2) is also illustrated with the computer routine FCHISQ of Program

Program 10-2 **FCHISQ** Reduced chi-square for fit χ_ν^2.

```
C      FUNCTION FCHISQ
C
C      PURPOSE
C        EVALUATE REDUCED CHI SQUARE FOR FIT TO DATA
C           FCHISQ = SUM ((Y-YFIT)**2 / SIGMA**2) / NFREE
C
C      USAGE
C        RESULT = FCHISQ (Y, SIGMAY, NPTS, NFREE, MODE, YFIT)
C
C      DESCRIPTION OF PARAMETERS
C        Y        - ARRAY OF DATA POINTS
C        SIGMAY   - ARRAY OF STANDARD DEVIATIONS FOR DATA POINTS
C        NPTS     - NUMBER OF DATA POINTS
C        NFREE    - NUMBER OF DEGREES OF FREEDOM
C        MODE     - DETERMINES METHOD OF WEIGHTING LEAST-SQUARES FIT
C                   +1 (INSTRUMENTAL) WEIGHT(I) = 1./SIGMAY(I)**2
C                    0 (NO WEIGHTING) WEIGHT(I) = 1.
C                   -1 (STATISTICAL)  WEIGHT(I) = 1./Y(I)
C        YFIT     - ARRAY OF CALCULATED VALUES OF Y
C
C      SUBROUTINES AND FUNCTION SUBPROGRAMS REQUIRED
C        NONE
C
C      MODIFICATIONS FOR FORTRAN II
C        OMIT DOUBLE PRECISION SPECIFICATIONS
C
       FUNCTION FCHISQ (Y, SIGMAY, NPTS, NFREE, MODE, YFIT)
       DOUBLE PRECISION CHISQ, WEIGHT
       DIMENSION Y(1), SIGMAY(1), YFIT(1)
    11 CHISQ = 0.
    12 IF (NFREE) 13, 13, 20
    13 FCHISQ = 0.
       GO TO 40
C
C          ACCUMULATE CHI SQUARE
C
    20 DO 30 I=1, NPTS
    21 IF (MODE) 22, 27, 29
    22 IF (Y(I)) 25, 27, 23
    23 WEIGHT = 1. / Y(I)
       GO TO 30
    25 WEIGHT = 1. / (-Y(I))
       GO TO 30
    27 WEIGHT = 1.
       GO TO 30
    29 WEIGHT = 1. / SIGMAY(I)**2
    30 CHISQ = CHISQ + WEIGHT*(Y(I)-YFIT(I))**2
C
C          DIVIDE BY NUMBER OF DEGREES OF FREEDOM
C
    31 FREE = NFREE
    32 FCHISQ = CHISQ / FREE
    40 RETURN
       END
```

10-2. This is a Fortran function subprogram to evaluate χ_ν^2 for a set of data points y_i and a fitted function $y(x_i)$. The input variables are Y, SIGMAY, NPTS, NFREE, MODE, and YFIT. The result is returned to the calling program as the value of the function FCHISQ.

As in previous programs, the data points y_i are assumed to be stored in the array Y, and the standard deviations of these points σ_i are assumed to be stored in the array SIGMAY. The variable NPTS $= N$ represents the number of data points, and the variable NFREE represents the number of degrees of freedom ν. Provision is included for examining this variable in statement 12 to make sure it is positive and nonzero. Otherwise the calculation is invalid and the result for χ_ν^2 is set equal to 0.

The variable MODE determines the method of weighting as in Programs 6-1, 8-1, and 8-2. For MODE > 0, the standard deviations σ_i are extracted from the array SIGMAY. For MODE < 0, they are estimated from the individual data points assuming a Poisson distribution $\sigma_i^2 = y_i$. For MODE $= 0$, the uncertainties σ_i are assumed to be unknown and a calculation of χ_ν^2 is not possible. What is returned to the calling program in this case is the estimated variance of the data s^2. The array YFIT is assumed to contain the values of the fitted function YFIT(I) $= y(x_i)$.

The value for χ^2 is accumulated in statement 30 using the weighting factors $1/\sigma_i^2$ evaluated in statements 21–29 as part of the DO loop over the N data points in statements 20–30. The reduced chi-square is evaluated in statement 32 and returned to the calling program.

This subprogram can, of course, be used to evaluate χ_ν^2 for any arbitrary function, regardless of whether or not it is linear in the coefficients. We will use it in this way in Chapter 11.

10-2 F TEST

As discussed in the previous section, the χ^2 test is somewhat ambiguous unless the form of the parent function is known because the statistic χ^2 measures not only the discrepancy between the estimated function and the parent function, but also the deviations between the data and the parent function simultaneously. We would prefer a test which separates these two types of information so that we can concentrate on the former type. One such test is the F test which combines two different methods of determining a χ^2 statistic and compares the results to see if their relation is reasonable.

F distribution If two statistics χ_1^2 and χ_2^2 are determined which follow the χ^2 distribution, the ratio of the reduced chi-squares $\chi_{\nu_1}^2$ and $\chi_{\nu_2}^2$ are distributed according to the F distribution[1] $P_f(f, \nu_1, \nu_2)$

$$f = \frac{\chi_{\nu_1}^2}{\chi_{\nu_2}^2} = \frac{\chi_1^2/\nu_1}{\chi_2^2/\nu_2} \tag{10-6}$$

$$P_f(f, \nu_1, \nu_2) = \frac{\Gamma[(\nu_1 + \nu_2)/2]}{\Gamma(\nu_1/2)\Gamma(\nu_2/2)} \left(\frac{\nu_1}{\nu_2}\right)^{\nu_1/2} \frac{f^{\frac{1}{2}(\nu_1 - 1)}}{(1 + f\nu_1/\nu_2)^{\frac{1}{2}(\nu_1 + \nu_2)}}$$

where ν_1 and ν_2 are the numbers of degrees of freedom corresponding to χ_1^2 and χ_2^2. By the definition of χ^2, a ratio of ratios of variances

$$\frac{\chi_{\nu_1}^2}{\chi_{\nu_2}^2} = \frac{s_1^2/\sigma_1^2}{s_2^2/\sigma_2^2}$$

is also distributed as F, where s_1 and s_2 are experimental estimates of standard deviations σ_1 and σ_2 pertaining to some characteristic of the same or different distributions.

As with our tests of χ^2 and the linear-correlation coefficient r, we will be more interested in the integral probability $P_F(F, \nu_1, \nu_2)$

$$P_F(F, \nu_1, \nu_2) = \int_F^\infty P_f(f, \nu_1, \nu_2) \, df$$

which describes the probability of observing such a large value of F from a random set of data compared with the correct fitting function. This function is tabulated and graphed in Table C-5 for a wide range of F, ν_1, and ν_2.

A word of caution is in order concerning the use of these tables. Since the statistic defined in Equation (10-6) on which the F test is made is defined as the ratio of two determinations of χ^2

[1] See Pugh and Winslow, sec. 12-7, for a derivation.

without specifying which must be in the numerator, we must be able to define two statistics F_{12} and F_{21}

$$F_{12} = \frac{\chi^2_{\nu_1}}{\chi^2_{\nu_2}} \qquad F_{21} = \frac{\chi^2_{\nu_2}}{\chi^2_{\nu_1}} = \frac{1}{F_{12}}$$

which must both be distributed according to the F distribution.

If in some experiment our calculations yield a particular value of F_{12}, we can use Table C-5 to determine whether such a large value is less than 5% probable (Table C-6 and Figure C-6) or less than 1% probable (Table C-7 and Figure C-7). If the test value is less than the tabulated values, we must also make sure that it is not too small. To do this we compare the value

$$F_{21} = 1/F_{12}$$

to the same tables and graphs, noting that the values of ν_1 and ν_2 are reversed. The values of ν_1 and ν_2 specified in Table C-5 correspond to the degrees of freedom for the numerator and denominator of Equation (10-6), respectively.

EXAMPLE 10-3   For example, suppose $F_{12} = 0.2$ with $\nu_1 = 2$ and $\nu_2 = 10$. For Table C-6, the observed value of F_{12} may be as high as 4.10 and still be exceeded by about 5% of random observations. Similarly, we compare $F_{21} = 1/F_{12} = 5.0$ with the 5% point for $\nu_1 = 10$ and $\nu_2 = 2$ which has a value of 19.4. Since the values of F_{12} and F_{21} are well within the 5% limits, we can have confidence in the fit.

What we are estimating in this example is the probability $P_F(F_{12}, \nu_1, \nu_2)$ that F_{12} is not too large and the probability $P_F(1/F_{12}, \nu_2, \nu_1)$ that F_{12} is not too small. It is tempting to simplify this procedure by assuming that

$$P_F(1/F_{12}, \nu_2, \nu_1) = P_F(1/F_{12}, \nu_1, \nu_2)$$

so that our test consists of determining F such that

$$P_F(F, \nu_1, \nu_2) = 0.05$$

and requiring that

$$F > F_{12} > 1/F$$

This approximation is valid for reasonably large values of ν_1 and ν_2 but not for small values of either (for example, see above where $4.10 > F_{12} > 1/19.4$).

Multiple-correlation coefficient   There are two types of F tests which are normally performed on least-squares fitting procedures. One is designed to test the entire fit and can be related to the multiple-correlation coefficient R. The other, to be discussed later, tests the inclusion of an additional term in the fitting function.

If we consider the sum of squares of deviations S_y^2 associated with the spread of the data points around their mean (omitting factors of $1/\sigma_i^2$ for clarity),

$$S_y^2 = \Sigma(y_i - \bar{y})^2$$

this is a statistic which follows the χ^2 distribution with $N - 1$ degrees of freedom (only one parameter $\bar{y}$ must be fitted to N data points). It is a characteristic of quantities which follow the χ^2 distribution that they may be expressed as the sum of other quantities which also follow the χ^2 distribution such that the number of degrees of freedom of the original statistic is the sum of the numbers of degrees of freedom of the terms in the sum.

By suitable manipulation and rearrangement, it can be shown that S_y^2 can be expressed as the sum of two terms.

$$\Sigma(y_i - \bar{y})^2 = \sum \left[(y_i - \bar{y}) \sum_{j=1}^{n} a_j(X_j - \bar{X}_j) \right]$$
$$+ \sum \left(y_i - \sum_{j=1}^{n} a_j X_j \right)^2$$
$$= \sum_{j=1}^{n} a_j \Sigma[(y_i - \bar{y})(X_j - \bar{X}_j)] + \Sigma[y_i - y(x_i)]^2$$

$$(10\text{-}7)$$

The left side of Equation (10-7) is distributed as χ^2 with $N - 1$ degrees of freedom. The right-hand term is our definition of χ^2 from Equation (10-2) and has $N - n - 1$ degrees of freedom.

Consequently, the middle term must be distributed according to the χ^2 distribution with n degrees of freedom.

By comparing this middle term with our definition of the multiple-correlation coefficient R in Equation (7-18), we can express it as a fraction R^2 of the statistic S_y^2.

$$\sum_{j=1}^{n} a_j \Sigma[(y_i - \bar{y})(X_j - \bar{X}_j)] = R^2 \Sigma(y_i - \bar{y})^2$$

Consequently, the right-hand term of Equation (10-7) must be equivalent to the remaining fraction $1 - R^2$ of the statistic S_y^2. Equation (10-7) becomes

$$\Sigma(y_i - \bar{y})^2 = R^2 \Sigma(y_i - \bar{y})^2 + (1 - R^2)\Sigma(y_i - \bar{y})^2 \qquad (10\text{-}8)$$

where as before both terms on the right-hand side are distributed as χ^2, the first with n degrees of freedom and the second with $N - n - 1$ degrees of freedom.

Thus, the physical meaning of the multiple-correlation coefficient becomes evident. It divides the total sum of squares of deviations S_y^2 into two parts. The first fraction $R^2 S_y^2$ is the sum of squares due to the regression and is a measure of the spread of the dependent and independent variable data in data space. The second fraction $(1 - R^2)S_y^2$ is the sum of squares of the deviations about the regression and represents the agreement between the fit and the data.

From the definition of Equation (10-6), we can define a ratio F_R of the terms in the right-hand side of Equation (10-8) which follows the F distribution with $\nu_1 = n$ and $\nu_2 = N - n - 1$ degrees of freedom.

$$F_R = \frac{R^2/n}{(1 - R^2)/(N - n - 1)} = \frac{R^2(N - n - 1)}{(1 - R^2)n} \qquad (10\text{-}9)$$

From this definition of F_R in terms of the multiple-correlation coefficient R, it is clear that a large value of F_R corresponds to a good fit, where the multiple correlation is good and $R \simeq 1$. The F test for this statistic is, in fact, a test that the coefficients are 0: $a_j = 0$. So long as F_R exceeds the test value for

F, we can be fairly confident that our coefficients are nonzero. If, on the other hand, $F_R < F$, we may conclude that at least one of the terms in the fitting function is not valid, is decreasing the multiple correlation by its inclusion, and should have a coefficient of 0.

EXAMPLE 10-4   The subroutine REGRES of Program 9-1, for example, contains the calculation for F_R of Equation (10-9). The value of R^2 is calculated in statement 133 as RMUL and used in the following statement to calculate FTEST = F_R where

$$\text{FREEJ} = \nu_1 = n$$

and FREEN = $\nu_2 = N - n - 1$. This value of F_R is not used within the subroutine but is returned to the main calling program where it may be used in conjunction with the variable JTERMS and the function subprogram FCTN to modify the selection of terms in the fitting function as described in Section 9-2.

Test of additional term   Because of the additivity nature of statistics which obey the χ^2 statistics, we can form a new χ^2 statistic by taking the difference of two other statistics which are distributed as χ^2. In particular, if we fit a set of data with a fitting function with $n - 1$ terms (plus a constant term), the resulting value of chi-square associated with the deviations about the regression $\chi^2(n - 1)$ has $N - n$ degrees of freedom. If we add another term to the fitting function, the corresponding value of chi-square $\chi^2(n)$ has $N - n - 1$ degrees of freedom. The difference between these two must follow the χ^2 distribution with 1 degree of freedom.

If we form the ratio of the difference $\chi^2(n - 1) - \chi^2(n)$ over the new value $\chi^2(n)$, we can form a statistic F_χ which follows the F distribution with $\nu_1 = 1$ and $\nu_2 = N - n - 1$.

$$F_\chi = \frac{\chi^2(n - 1) - \chi^2(n)}{\chi^2(n)/(N - n - 1)} \tag{10-10}$$

This ratio is a measure of how much the additional term has

improved the value of the reduced chi-square

$$F_\chi = \frac{\Delta\chi^2}{\chi^2_\nu}$$

and should be small when the function with n terms does not significantly improve the fit over the function with $n - 1$ terms. Thus, we can be confident in the relative merit of new terms if the value of F_χ is large. As for F_R, this is really a test of whether the coefficient for the new term is 0: $a_n = 0$. If $F_\chi > F$, we can be fairly confident the coefficient should not be 0 and the term, therefore, should be included.

Table C-5 and Figure C-5 are useful for testing F_χ. They give the value of F corresponding to various values of the probability $P_F(F,1,\nu_2)$ and various values of ν_2 for the case where $\nu_1 = 1$. Thus, rather than evaluating F for critical values of the probability (for example, 5% or 1%), we can evaluate the probability corresponding to the observed value F_χ.

EXAMPLE 10-5 The use of this F_χ test is illustrated in the subroutine LEGFIT of Program 8-2. In statement 141, the value of FVALUE $= F_\chi$ is evaluated according to Equation (10-10) where CHISQ1 is the value of $\chi^2(n - 1)$ saved from the previous iteration of the number of terms, CHISQ $= \chi^2(n)$ is the new value, and FREE $= N - n$ is the number of degrees of freedom (note that the constant term is included in n). In the following statement, the observed value of F_χ is compared with test values of F which are supplied by the main program. If $F_\chi > F$, the program branches to statement 134 and increases the number of terms NTERMS $= n$ in the fitting function by 1 (provided the maximum number JMAX is not exceeded). The value of CHISQ $= \chi^2(n)$ is saved in CHISQ1 for use in testing the next iteration, and the program returns to statement 51 to fit the data with $n + 1$ terms in the fitting function.

If $F_\chi < F$, the assumption is that the last term added to the fitting function is not justified and the final result must be a fit to the function with $n - 1$ terms. The program branches to statement 143 where the number of terms is decreased by 1 and the program returns to statement 51 to accumulate the fit to $n - 1$ terms.

The value of JMAX is adjusted so that the test of statement 131 will cause the subroutine to branch to statement 151 and return to the main program.

In this manner we can use the F_χ test to test the validity of each term as it is added to the fitting function. When the F_χ test fails, the function is assumed to be too complex to describe the data. Higher-order polynomials are presumably not justified by the least-squares fitting.

This method of cutting off the fitting function may result in an improper fit if the appropriate function contains some but not all the terms up to a maximum order. The cutoff will occur when the first noncorrelated term is attempted without testing whether higher-order terms are justified. For this reason, the test of F_χ should generally be accompanied by some test of the overall fit, such as a test of F_R or of χ^2 to see if even higher-order terms should be investigated.

SUMMARY

Goodness-of-fit criterion χ^2:

$$\chi^2 = \sum \left\{ \frac{1}{\sigma_i{}^2} [y_i - y(x_i)]^2 \right\}$$

Degrees of freedom: $\nu = N - n - 1$ for a fit to a function with n coefficients plus one constant term.

Reduced chi-square:

$$\chi_\nu^2 = \frac{\chi^2}{\nu}$$

Probability $P_\chi(\chi^2,\nu)$ that any random set of N data points would yield a value of chi-square as large as or larger than χ^2:

$$P_\chi(\chi^2,\nu) = \int_{\chi^2}^{\infty} \frac{z^{\frac{1}{2}(\nu-2)}e^{-z/2}}{2^{\nu/2}\Gamma(\nu/2)} \, dz = e^{-\chi^2/2} \sum_{m=0}^{\frac{1}{2}(\nu-2)} \frac{(\chi^2/2)^m}{m!} \qquad \nu \text{ even}$$

F test:

$$f = \frac{\chi^2_{\nu_1}}{\chi^2_{\nu_2}}$$

$$P_F(F,\nu_1,\nu_2) = \int_F^\infty P_f(f,\nu_1,\nu_2)\, df$$

F test for multiple-correlation coefficient R (for $\nu = N - n - 1$):

$$F_R = \frac{R^2/n}{(1 - R^2)/(N - n - 1)} = \frac{R^2(N - n - 1)}{(1 - R^2)n}$$

F test for χ^2 for validity of adding nth term (plus constant term):

$$F_\chi = \frac{\chi^2(n - 1) - \chi^2(n)}{\chi^2(n)/(N - n - 1)} = \frac{\Delta\chi^2}{\chi^2_\nu}$$

EXERCISES

10-1 For a typical number of degrees of freedom ($\nu \simeq 10$), what is the range of probability $P_\chi(\chi^2,\nu)$ of finding χ^2_ν as small as 0.5 or as large as 1.5?

10-2 Discuss the meaning of χ^2 and justify the relationship between it and the sample variance $s^2 = \chi^2_\nu$.

10-3 What is the probability of finding a value of $\chi^2_\nu = 1.5$ with $\nu = 100$ degrees of freedom? Would you consider this a reasonably good fit?

10-4 Is a large value of F good or bad?

10-5 If we wish to set an arbitrary criterion of a probability of 0.01 for the F_χ test, what would be a reasonable average value for F test?

10-6 What different aspects of a fit do the F_R and F_χ tests represent?

CHAPTER ELEVEN

LEAST-SQUARES FIT TO
AN ARBITRARY FUNCTION

The methods of least squares and multiple regression developed in Chapters 6–9 are restricted to fitting functions which are linear in the coefficients as in Equation (9-1).

$$y(x) = a_0 + \sum_{j=1}^{n} [a_j X_j(x)]$$

Let us extrapolate these methods to develop a technique for fitting the data y_i with a function $y(x)$ which is not linear in its parameters, i.e., the method of *non-linear least squares*.

Consider the function $y(x)$ with parameters a_j. For example,

$y(x)$ could be a transcendental function

$$y(x) = a_1 \sin (a_2 x)$$

or a Gaussian peak plus a quadratic background

$$y(x) = a_1 \exp \left[-\frac{1}{2} \left(\frac{x - a_2}{a_1} \right)^2 \right] + a_4 + a_5 x + a_6 x^2 \quad (11\text{-}1)$$

or any other function such that some of the parameters cannot be separated into different terms of a sum.

As before, we can define a measure of goodness of fit χ^2

$$\chi^2 \equiv \sum \left\{ \frac{1}{\sigma_i^2} [y_i - y(x_i)]^2 \right\} \quad (11\text{-}2)$$

where the σ_i are the uncertainties in the data points y_i. There are three sources of error which contribute to the size of χ^2:

1. The data y_i are a random sample from the parent population with expected values $\langle y_i \rangle$ given by the parent distribution. The fluctuations of the y_i about the expected values $\langle y_i \rangle$ may be statistically greater or less than the expected uncertainty σ_i.
2. χ^2 is a continuous function of all parameters a_j.
3. The choice of the functional behavior of the analytical function $y(x)$ as an approximation to the "true" function $\langle y(x) \rangle$ will influence the range of possible values for χ^2.

Nothing can be done about the contributions from (1) without repeating the experiment. The optimum values for the parameters a_j can be estimated with the least-squares method by minimizing the contributions from (2). The resultant value of χ^2 for several different functions $y(x)$ can be compared to determine the most probable functional form for $y(x)$ as in (3).

Method of least squares According to the method of least squares, the optimum values of the parameters a_j are

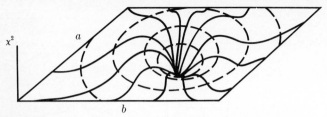

FIGURE 11-1 Representative hypersurface describing variation of χ^2 vs. two parameters a and b.

obtained by minimizing χ^2 with respect to each of the parameters simultaneously.

$$\frac{\partial}{\partial a_j} \chi^2 = \frac{\partial}{\partial a_j} \sum \left\{ \frac{1}{\sigma_i^2} [y_i - y(x_i)]^2 \right\} = 0$$

It is generally not convenient to derive an analytical expression for calculating the parameters of a non-linear function $y(x)$. Instead, χ^2 must be considered a continuous function of the n parameters a_j describing a hypersurface in n-dimensional space as illustrated in Figure 11-1, and the space must be searched for the appropriate minimum value of χ^2.

One of the difficulties of such a search is that for an arbitrary function there may be more than one local minimum for χ^2 within a reasonable range of values for the parameters a_j. Unless the range can be restricted to a region in which it is known that there is only one minimum, it may be advantageous to conduct a coarse grid mapping of the parameter space to locate the main minima and identify the desired range of parameters over which to refine the search.

In the simplest brute-force mapping procedure, the permissible range for each parameter a_j is divided into n equal increments Δa_j so that the n-parameter space is divided into $\prod_{j=1}^{n} n_j$ hypercubes. The value of χ^2 is then evaluated at each of the vertices of these hypercubes. This procedure yields a coarse map of the behavior of χ^2 as a function of all of the parameters a_j.

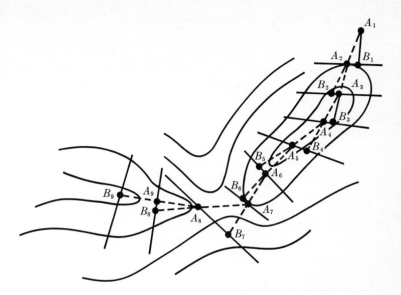

FIGURE 11-2 Ravine search path. (*Reprinted by permission from Arndt and MacGregor, p. 268.*)

Alternatively, the methods of searching parameter space described in Section 11-2 can be used with the search procedure started at several different points in the parameter space. If enough different starting points are chosen, all of the important local minima should be discovered by this method.

Ravine search A more sophisticated method of locating the various minima of the χ^2 hypersurface[1] involves traversing the surface from minimum to minimum by the path of lowest value in χ^2, as a river follows a ravine in traversing from lake to lake. Consider, for example, the contour plot of χ^2 for a two-parameter function in Figure 11-2. The solid curves indicate the points of constant χ^2. Concentric curves indicate variations in χ^2 by equal increments. Starting at point A_1, the search traverses the length of the first minimum, then continues along the same gen-

[1] See Tyapkin, p. 138, and Arndt and MacGregor, pp. 268–269.

eral direction (within 90°) but in the direction that minimizes the new value of χ^2.

From point A_1, a vector of arbitrary length λ is drawn in the direction $-\gamma_1$ of decreasing χ^2 to the point $B_1 = A_1 - \lambda\gamma_1$ (see the discussion of the gradient γ in Section 11-2). A new direction $-\gamma_2$ of decreasing χ^2 is determined from point B_1, and the point A_2 is located at the point of minimum χ^2 along this vector by assuming a parabolic behavior for χ^2 (see the discussion of parabolic interpolation in Section 11-2). The point B_2 is determined by extrapolating a vector from A_1 through A_2 by a length λ. The direction $-\gamma_3$ of decreasing χ^2 is determined at point B_2, and the point A_3 is located at the minimum of χ^2 along this vector.

The search continues in this manner, extrapolating by a length λ along the vectors through the points A_j to the points B_j, and then locating new points A_{j+1} at nearby local minima. The extrapolation forces the search to continue in the same general direction, and the local minimization forces the search to follow the ravines of the χ^2 hypersurface.

11-2 SEARCHING PARAMETER SPACE

The method of least squares consists of determining the values of the parameters a_j of the function $y(x)$ which yield a minimum for the function χ^2 given in Equation (11-2). There are a number of ways of finding this minimum value. Methods of searching parameter space will be described in this section, and approximate analytical methods will be described in Sections 11-3 and 11-4.

Grid search If the variation of χ^2 with each parameter a_j is fairly independent of how well optimized the other parameters are, then the optimum values can be determined most simply by minimizing χ^2 with respect to each parameter separately. This is the method of the *grid search*. With successive iterations of locating the local minimum for each parameter in turn, the absolute minimum may be located with any desired precision. The main disadvantage is that if the variations of χ^2 with various param-

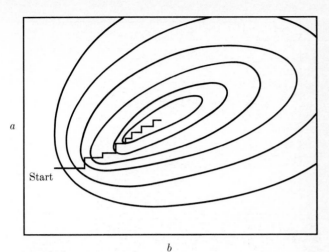

a

Start

b

FIGURE 11-3 Tortuous path of grid search in two-parameter space.

eters are not independent, then this method may converge very slowly toward the minimum.

Consider, for example, the contour plot of χ^2 for two parameters in Figure 11-3. The variation of χ^2 is generally approximately elliptical near the minimum, and, typically, the ellipse is elongated as in Figure 11-3. If a grid search is initiated near one end of the ellipse, the grid search is very inefficient, as indicated by the solid line zigzagging across the ellipse. Nevertheless, the simplicity of the calculations involved in a grid search often compensate for this inefficiency.

The procedure of the grid search is as follows:

1. One parameter a_j is incremented by a quantity Δa_j, where the magnitude of this quantity is specified and the sign is chosen such that χ^2 decreases.
2. The parameter a_j is repeatedly incremented by the same amount Δa_j until χ^2 starts to increase.
3. Assuming the variation of χ^2 near the minimum can be described in terms of a parabolic function of the parameter a_j, we can use the values of χ^2 for the last three values of a_j to

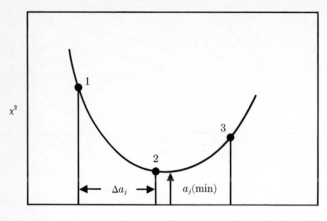

FIGURE 11-4 Parabolic interpolation to find position $a_j(min)$ for minimum χ^2 using the values of χ^2 for $a_j(1)$, $a_j(2)$, and $a_j(3)$.

determine the minimum of the parabola in Figure 11-4.

$$a_j(3) = a_j(2) + \Delta a_j = a_j(1) + 2\Delta a_j$$
$$\chi^2(3) > \chi^2(2) \leq \chi^2(1) \tag{11-3}$$

4. The minimum of the parabola is given by

$$a_j(min) = a_j(3) - \Delta a_j \left[\frac{\chi^2(3) - \chi^2(2)}{\chi^2(3) - 2\chi^2(2) + \chi^2(1)} + \frac{1}{2} \right] \tag{11-4}$$

5. χ^2 is minimized for each parameter in turn.
6. The above procedure is repeated until the last iteration yields a negligibly small decrease in χ^2.

Program 11-1 The grid-search method is illustrated in the computer routine GRIDLS of Program 11-1. This is a Fortran subroutine to optimize the parameters a_j of a function $y(x)$ by the grid-search method of least squares. The input variables are X, Y, SIGMAY, NPTS, NTERMS, MODE, A, and DELTAA, and the output variables are A, SIGMAA, YFIT, and CHISQR.

As in Program 9-1, the values of the independent variable x_i

are assumed to be stored in the array x, and the data for the dependent variable y_i are assumed to be stored in the array y with identical ordering. The variable NPTS $= N$ represents the number of pairs of data points (x_i, y_i), and the variable NTERMS $= n$ represents the number of parameters a_j in the fitting function $y(x)$. The variable MODE determines the method of weighting the fit with the uncertainties SIGMAY $= \sigma_i$ as in Program 9-1. Array A must contain starting values of parameters a_j and array DELTAA must contain the values of step sizes Δa_j by which the parameters a_j are to be incremented during the search.

The variable NFREE $= \nu = N - n$, evaluated in statement 11, is the number of degrees of freedom in the fit and must be greater than 0 for a legitimate least-squares fit. The function YFIT(I) $= y(x_I)$ and the value of the reduced chi-square CHISQ1 $= \chi_\nu^2$ for the initial value of the parameter a_j are evaluated in statements 21–23. The function $y(x)$ must be supplied by an additional function subprogram FUNCTN(X,I,A) $= y(x_i)$ similar to the linear function FCTN(X,I,J,M) for Program 9-1. This routine will be described in Program 11-2. The value of χ^2 is evaluated with the function subprogram FCHISQ of Program 10-1.

One parameter A(J) $= a_J$ is incremented by an amount DELTAA(J) $= \Delta a_J$ in statements 41–43 with the sign chosen in statements 45–57 so that χ^2 decreases. When the value of χ^2 begins to increase after successive increments of a_J, as discovered in statement 66, so that the conditions of Equations (11-3) are satisfied, the parabolic interpolation of χ^2 in Equation (11-4) is used to determine the value of a local minimum for χ^2 (as a function of a_J) in statements 81–82.

The DO loop of statements 20–90 repeats the above procedure for each of the parameters a_J in turn. The final values of

$$\text{YFIT(I)} = y(x_I)$$

and CHISQR $= \chi_\nu^2$ are evaluated in statements 91–93 before returning to the main program. After each iteration, the parameter increments DELTAA(J) $= \Delta a_J$ are modified in statement 84 to optimize the speed of the search. If the search requires five determinations of χ^2 at five successive values of a parameter a_j, the

Program 11-1   GRIDLS   Grid-search least-squares fit for a nonlinear function.

```
C       SUBROUTINE GRIDLS
C
C
C       PURPOSE
C         MAKE A GRID-SEARCH LEAST-SQUARES FIT TO DATA WITH A SPECIFIED
C             FUNCTION WHICH IS NOT LINEAR IN COEFFICIENTS
C
C       USAGE
C         CALL GRIDLS (X, Y, SIGMAY, NPTS, NTERMS, MODE, A, DELTAA,
C             SIGMAA, YFIT, CHISQR)
C
C       DESCRIPTION OF PARAMETERS
C         X       - ARRAY OF DATA POINTS FOR INDEPENDENT VARIABLE
C         Y       - ARRAY OF DATA POINTS FOR DEPENDENT VARIABLE
C         SIGMAY  - ARRAY OF STANDARD DEVIATIONS FOR Y DATA POINTS
C         NPTS    - NUMBER OF PAIRS OF DATA POINTS
C         NTERMS  - NUMBER OF PARAMETERS
C         MODE    - DETERMINES METHOD OF WEIGHTING LEAST-SQUARES FIT
C                   +1 (INSTRUMENTAL) WEIGHT(I) = 1./SIGMAY(I)**2
C                    0 (NO WEIGHTING) WEIGHT(I) = 1.
C                   -1 (STATISTICAL)  WEIGHT(I) = 1./Y(I)
C         A       - ARRAY OF PARAMETERS
C         DELTAA  - ARRAY OF INCREMENTS FOR PARAMETERS A
C         SIGMAA  - ARRAY OF STANDARD DEVIATIONS FOR PARAMETERS A
C         YFIT    - ARRAY OF CALCULATED VALUES OF Y
C         CHISQR  - REDUCED CHI SQUARE FOR FIT
C
C       SUBROUTINES AND FUNCTION SUBPROGRAMS REQUIRED
C         FUNCTN (X, I, A)
C             EVALUATES THE FITTING FUNCTION FOR THE ITH TERM
C         FCHISQ (Y, SIGMAY, NPTS, NFREE, MODE, YFIT)
C             EVALUATES REDUCED CHI SQUARE FOR FIT TO DATA
C
C       MODIFICATIONS FOR FORTRAN II
C         ADD F SUFFIX TO SQRT IN STATEMENT 83
C
C       COMMENTS
C         DELTAA VALUES ARE MODIFIED BY THE PROGRAM
C
```

increment is unchanged. If the search requires fewer points, the increment is reduced; otherwise it is increased according to the number of increments required.

One pass through this subroutine represents a single zigzag along the path of Figure 11-3. In practice, the search should be repeated until the value of CHISQR $= \chi_\nu^2$ does not change by more than an acceptable amount (for example, 1%). The initial values of DELTAA(J) $= \Delta a_j$ should be chosen to approach the minimum in only a few steps. As the search converges to the minimum, the values of DELTAA will be modified to permit a finer examination of the minimum and a consequently higher precision in locating it.

The calculation of the uncertainties σ_{a_j} in the parameters will be discussed in Section 11-5.

```
      SUBROUTINE GRIDLS (X, Y, SIGMAY, NPTS, NTERMS, MODE, A, DELTAA,
     1 SIGMAA, YFIT, CHISQR)
      DIMENSION X(1), Y(1), SIGMAY(1), A(1), DELTAA(1), SIGMAA(1),
     1 YFIT(1)
   11 NFREE = NPTS - NTERMS
      FREE = NFREE
      CHISQR = 0.
      IF (NFREE) 100, 100, 20
   20 DO 90 J=1, NTERMS
C
C         EVALUATE CHI SQUARE AT FIRST TWO SEARCH POINTS
C
   21 DO 22 I=1, NPTS
   22 YFIT(I) = FUNCTN (X, I, A)
   23 CHISQ1 = FCHISQ (Y, SIGMAY, NPTS, NFREE, MODE, YFIT)
      FN = 0.
      DELTA = DELTAA(J)
   41 A(J) = A(J) + DELTA
      DO 43 I=1, NPTS
   43 YFIT(I) = FUNCTN (X, I, A)
   44 CHISQ2 = FCHISQ (Y, SIGMAY, NPTS, NFREE, MODE, YFIT)
   45 IF (CHISQ1 - CHISQ2) 51, 41, 61
C
C         REVERSE DIRECTION OF SEARCH IF CHI SQUARE IS INCREASING
C
   51 DELTA = -DELTA
      A(J) = A(J) + DELTA
      DO 54 I=1, NPTS
   54 YFIT(I) = FUNCTN (X, I, A)
      SAVE = CHISQ1
      CHISQ1 = CHISQ2
   57 CHISQ2 = SAVE
C
C         INCREMENT A(J) UNTIL CHI SQUARE INCREASES
C
   61 FN = FN + 1.
      A(J) = A(J) + DELTA
      DO 64 I=1, NPTS
   64 YFIT(I) = FUNCTN (X, I, A)
      CHISQ3 = FCHISQ (Y, SIGMAY, NPTS, NFREE, MODE, YFIT)
   66 IF (CHISQ3 - CHISQ2) 71, 81, 81
   71 CHISQ1 = CHISQ2
      CHISQ2 = CHISQ3
      GO TO 61
C
C         FIND MINIMUM OF PARABOLA DEFINED BY LAST THREE POINTS
C
   81 DELTA = DELTA * (1./(1.+(CHISQ1-CHISQ2)/(CHISQ3-CHISQ2)) + 0.5)
   82 A(J) = A(J) - DELTA
   83 SIGMAA(J) = DELTAA(J) * SQRT (2./(FREE*(CHISQ3-2.*CHISQ2+CHISQ1)))
   84 DELTAA(J) = DELTAA(J) * FN/3.
   90 CONTINUE
C
C         EVALUATE FIT AND CHI SQUARE FOR FINAL PARAMETERS
C
   91 DO 92 I=1, NPTS
   92 YFIT(I) = FUNCTN (X, I, A)
   93 CHISQR = FCHISQ (Y, SIGMAY, NPTS, NFREE, MODE, YFIT)
  100 RETURN
      END
```

Program 11-2 **FUNCTN** (Gaussian + quadratic) Non-linear fitting function.

```
C      FUNCTION FUNCTN (GAUSSIAN + QUADRATIC)
C
C      PURPOSE
C        EVALUATE TERMS OF FUNCTION FOR NON-LINEAR LEAST-SQUARES SEARCH
C          WITH FORM OF A GAUSSIAN PEAK PLUS QUADRATIC POLYNOMIAL
C          FUNCTN(X,I,A) = A(1)*EXP(-Z**2/2) + A(4) + A(5)*X + A(6)*X**2
C          WHERE X = X(I) AND Z = (X - A(2))/A(3)
C
C      USAGE
C        RESULT = FUNCTN (X, I, A)
C
C      DESCRIPTION OF PARAMETERS
C        X      - ARRAY OF DATA POINTS FOR INDEPENDENT VARIABLE
C        I      - INDEX OF DATA POINTS
C        A      - ARRAY OF PARAMETERS
C
C      SUBROUTINES AND FUNCTION SUBPROGRAMS REQUIRED
C        NONE
C
C      MODIFICATIONS FOR FORTRAN II
C        ADD F SUFFIX TO EXP IN STATEMENT 16
C
       FUNCTION FUNCTN (X, I, A)
       DIMENSION X(1), A(1)
11     XI = X(I)
12     FUNCTN = A(4) + A(5)*XI + A(6)*XI**2
13     Z = (XI - A(2)) / A(3)
       Z2 = Z**2
       IF (Z2 - 50.) 16, 20, 20
16     FUNCTN = FUNCTN + A(1)*EXP (-Z2/2.)
20     RETURN
       END
```

Program 11-2 The computer routine FUNCTN of Program 11-2 illustrates the use of such a function in the program GRIDLS. This is a Fortran function subprogram designated FUNCTN for compatibility with GRIDLS which evaluates the Gaussian function plus a quadratic polynomial of Equation (11-1). The input variables X, I, and A are defined for compatibility with Program 11-1. The independent variables x_i are assumed to be stored in the array X, and the six parameters a_j are assumed to be stored in the array A. The variable I = i is the index of the event XI = x_i. The dimensionless deviation Z of the Gaussian function is evaluated in statement 13, and the function $y(x)$ of Equation (11-1) is evaluated in statements 12 and 16 and returned to the calling subroutine as the value of the function FUNCTN.

EXAMPLE 11-1 This particular choice of the function $y(x)$ would be appropriate for fitting a Gaussian peak on a slowly

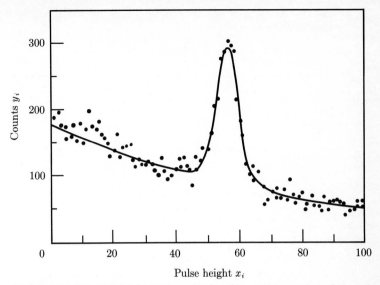

FIGURE 11-5 Multichannel pulse-height spectrum for data of Example 11-1. Solid curve is fit to data with a Gaussian peak plus a quadratic-polynomial background.

varying background like that of Figure 11-5 which is part of a multichannel pulse-height spectrum giving the number of counts in a detector y_i vs. the channel number x_i corresponding to the energy recorded by the detector for each count. Table 11-1 gives the starting values for the six parameters a_j and the values of these parameters at the end of each iteration of the search for the increments Δa_j specified. The search would approach the final value faster near the beginning of the search for larger values of the increments Δa_j but would converge to the actual minimum more slowly.

Gradient search The search could be improved if the zigzagging direction of travel indicated in Figure 11-3 were replaced with a more direct vector toward the appropriate minimum. In the *gradient-search* method of least squares, all the parameters a_j are incremented simultaneously, with the relative

Table 11-1 Search paths for non-linear least-squares fitting procedures. Fitting function is

$$y(x) = a_1 \exp\left[-\frac{1}{2}\left(\frac{x - a_2}{a_3}\right)^2 \right] + a_4 + a_5 x + a_6 x^2$$

Routine	χ_ν^2	a_1	a_2	a_3	a_4	a_5	a_6
GRIDLS	16.1	200.0	56.00	4.000	200.0	-2.000	.01000
	3.991	138.5	55.89	3.153	169.6	-2.066	.00962
	1.807	230.0	56.03	3.460	171.2	-2.109	.00946
	1.586	219.8	56.02	3.611	174.5	-2.141	.00935
	1.459	212.0	56.02	3.653	177.1	-2.165	.00926
	1.3933	208.7	56.03	3.649	179.1	-2.183	.00920
	1.3563	207.6	56.03	3.632	180.6	-2.197	.00916
	1.3350	207.3	56.04	3.612	181.8	-2.208	.00912
	1.3227	207.3	56.04	3.595	182.7	-2.216	.00910
	1.3156	207.4	56.04	3.581	183.3	-2.223	.00908
	1.3114	207.6	56.04	3.571	183.8	-2.228	.00907
	1.3090	207.8	56.05	3.556	184.5	-2.231	.00906
	1.3067	207.9	56.05	3.562	184.5	-2.234	.00906
	1.3062	207.9	56.05	3.548	184.9	-2.239	.00906
	1.3058	208.0	56.05	3.546	185.0	-2.240	.00906
	1.3056	208.0	56.05	3.544	185.1	-2.242	.00906
	1.3055	208.0	56.05	3.542	185.2	-2.243	.00906
GRADLS	16.1	200.0	56.00	4.000	200.0	-2.000	.01000
	1.563	199.5	56.00	3.907	191.0	-2.250	.00803
	1.3218	198.6	56.03	3.616	185.6	-2.236	.00878
	1.3208	198.7	56.03	3.617	185.7	-2.234	.00890
	1.3206	198.7	56.03	3.617	185.6	-2.235	.00889
	1.3181	199.7	56.05	3.630	185.5	-2.235	.00897
	1.3170	199.7	56.05	3.629	185.4	-2.237	.00896
	1.3169	199.7	56.05	3.628	185.4	-2.238	.00896
CHIFIT	16.1	200.0	56.00	4.000	200.0	-2.000	.01000
	5.169	91.8	55.91	5.820	195.0	-2.547	.01297
	2.785	143.1	56.10	3.514	191.9	-2.421	.01142
	1.3526	198.9	56.00	3.848	188.4	-2.419	.01074
	1.3042	208.4	56.05	3.548	186.7	-2.307	.00961
	1.3041	208.2	56.05	3.553	186.6	-2.303	.00956
	1.3041	208.2	56.05	3.553	186.6	-2.303	.00956
CURFIT	16.1	200.0	56.00	4.000	200.0	-2.000	.01000
	1.3070	204.4	56.05	3.614	186.3	-2.288	.00944
	1.3041	207.9	56.05	3.563	186.6	-2.307	.00960
	1.3041	208.2	56.05	3.555	186.6	-2.304	.00957

magnitudes adjusted so that the resultant direction of travel in parameter space is along the gradient (or direction of maximum variation) of χ^2.

The gradient $\nabla\chi^2$, or direction in which χ^2 increases most rapidly, is a vector whose components are equal to the rate at which χ^2 increases in that direction.

$$\nabla\chi^2 = \sum_{j=1}^{n} \left[\frac{\partial\chi^2}{\partial a_j} \hat{a}_j \right]$$

(The hat over $\hat{a}_j$ indicates a unit vector.) In order to determine the gradient, the variation of χ^2 in the neighborhood of the starting point is sampled independently for each parameter to yield an approximate value for the first derivative.

$$(\nabla\chi^2)_j = \frac{\partial\chi^2}{\partial a_j} \simeq \frac{\chi^2(a_j + f\,\Delta a_j) - \chi^2(a_j)}{f\,\Delta a_j} \tag{11-5}$$

The amount by which a_j is changed in order to determine this derivative should be smaller than the step size Δa_j. The fraction f should be on the order of 10% ($f = 0.1$).

With this definition, the gradient has both magnitude and dimensions. In fact, if the dimensions of the various parameters a_j are not all the same, the components of the gradient do not even have the same dimensions. Let us define dimensionless parameters b_j by normalizing each of the parameters a_j to a size constant Δa_j which characterizes the variation of χ^2 with a_j rather roughly.

$$b_j = \frac{a_j}{\Delta a_j} \tag{11-6}$$

That is, we might use the values of Δa_j which were chosen for step sizes for increments in the grid search. We can then define a dimensionless gradient γ with a magnitude of unity.

$$\gamma_j = \frac{\partial\chi^2/\partial b_j}{\sqrt{\sum_{k=1}^{n} \left(\frac{\partial\chi^2}{\partial b_k}\right)^2}} \qquad \frac{\partial\chi^2}{\partial b_j} = \frac{\partial\chi^2}{\partial a_j} \Delta a_j \tag{11-7}$$

The direction which the gradient-search method follows is the *direction of steepest descent*, which is the opposite direction from the gradient γ. The search begins by incrementing all the parameters simultaneously by an amount δa_j whose relative value is given by the corresponding component of the dimensionless gradient γ, and whose absolute magnitude is given by the size constant Δa_j.

$$\delta a_j = -\gamma_j \, \Delta a_j \tag{11-8}$$

The minus sign ensures that the value of χ^2 decreases. The size constant Δa_j of Equation (11-8) must, of course, be the same as that of Equation (11-6).

There are several choices for methods of continuing the gradient search. The most straightforward is to recompute the gradient after each change in the parameters. One disadvantage of this method is that it is difficult to approach the bottom of the minimum asymptotically or even to identify that the solution is as close as it will get. Another is that recomputation of the gradient at each step for small step sizes results in an inefficient search, but the use of larger step sizes makes location of the minimum that less precise.

A reasonable perturbation on the method is to search along the direction of the original gradient in small steps, calculating only the value of χ^2 until the value of χ^2 begins to rise again. At this point, the gradient is recomputed and the search begins again in the new direction. Whenever the search straddles a minimum, a parabolic interpolation of χ^2 will improve the precision of locating the bottom of the minimum.

A more sophisticated approach is to use second partial derivatives of χ^2 calculated as in Equation (11-5) with finite differences to determine modifications to the gradient along the search path.

$$\frac{\partial \chi^2}{\partial a_j}\bigg|_{a_j + \delta a_j} \simeq \frac{\partial \chi^2}{\partial a_j}\bigg|_{a_j} + \sum_{k=1}^{n} \left(\frac{\partial^2 \chi^2}{\partial a_j \, \partial a_k} \, \delta a_k \right)$$

If the search is already fairly near the minimum, this method does decrease the number of steps needed, but at the expense of more elaborate computation. If the search is not near enough to the minimum, this method can actually increase the number of steps required if first-order perturbations on the gradient are not valid. For this reason we will ignore this type of approach. For further discussion, see Melkanoff et al.

The gradient search suffers markedly as the search approaches the minimum because the evaluation of the derivatives according to the method of Equation (11-5) consists of taking differences between nearly equal numbers. In fact, at the minimum of χ^2, these differences should vanish. For this reason, one of the methods discussed in the following sections may be used to locate the actual minimum once the gradient search has approached it fairly closely.

Program 11-3 The gradient-search method is illustrated in the computer routine GRADLS of Program 11-3. This is a Fortran subroutine to optimize the parameters a_j of a function $y(x)$ by the gradient-search method of least squares. The input variables are X, Y, SIGMAY, NPTS, NTERMS, MODE, A, and DELTAA, and the output variables are A, YFIT, and CHISQR. See the discussion of Program 11-1 for a description of these variables and of the function FUNCTN which must be supplied in addition.

The value of χ^2 at the start of the search is denoted CHISQ1. The variation of χ^2 in the immediate neighborhood is sampled in statements 32–39 according to Equation (11-5) with $f = 0.1$ to estimate the gradient components γ_j of Equation (11-7). The searching vector GRAD(J) = δa_J of Equation (11-8) is evaluated in statements 41–42.

The size of the step Δa_j is tested in statements 51–61. If the value of χ^2 at $a + \delta a$ is larger than that at a, the magnitudes of all the step sizes Δa_j are cut in half and the gradients recomputed. The value of χ^2 at the first search point $a_j + 0.1\delta a_j$ is denoted CHISQ2. The parameters a_j are then incremented by $0.1\delta a_j$ in statements 71–75 until the value of χ^2 starts to increase, as noted in statement 76. After each increment, if χ^2 has not increased, the previous values of χ^2 are relabelled in statements 81–82 so that

Program 11-3 GRADLS Gradient-search least-squares fit for a non-linear function.

```
C      SUBROUTINE GRADLS
C
C      PURPOSE
C        MAKE A GRADIENT-SEARCH LEAST-SQUARES FIT TO DATA WITH A
C            SPECIFIED FUNCTION WHICH IS NOT LINEAR IN COEFFICIENTS
C
C      USAGE
C        CALL GRADLS (X, Y, SIGMAY, NPTS, NTERMS, MODE, A, DELTAA,
C            YFIT, CHISQR)
C
C      DESCRIPTION OF PARAMETERS
C        X      - ARRAY OF DATA POINTS FOR INDEPENDENT VARIABLE
C        Y      - ARRAY OF DATA POINTS FOR DEPENDENT VARIABLE
C        SIGMAY - ARRAY OF STANDARD DEVIATIONS FOR Y DATA POINTS
C        NPTS   - NUMBER OF PAIRS OF DATA POINTS
C        NTERMS - NUMBER OF PARAMETERS
C        MODE   - DETERMINES METHOD OF WEIGHTING LEAST-SQUARES FIT
C                 +1 (INSTRUMENTAL) WEIGHT(I) = 1./SIGMAY(I)**2
C                  0 (NO WEIGHTING) WEIGHT(I) = 1.
C                 -1 (STATISTICAL)  WEIGHT(I) = 1./Y(I)
C        A      - ARRAY OF PARAMETERS
C        DELTAA - ARRAY OF INCREMENTS FOR PARAMETERS A
C        YFIT   - ARRAY OF CALCULATED VALUES OF Y
C        CHISQR - REDUCED CHI SQUARE FOR FIT
C
C      SUBROUTINES AND FUNCTION SUBPROGRAMS REQUIRED
C        FUNCTN (X, I, A)
C            EVALUATES THE FITTING FUNCTION FOR THE ITH TERM
C        FCHISQ (Y, SIGMAY, NPTS, NFREE, MODE, YFIT)
C            EVALUATES REDUCED CHI SQUARE FOR FIT TO DATA
C
C      MODIFICATIONS FOR FORTRAN II
C        ADD F SUFFIX TO SQRT IN STATEMENT 42
C
C      COMMENTS
C        DIMENSION STATEMENT VALID FOR NTERMS UP TO 10
C
```

CHISQ1, CHISQ2, and CHISQ3 are always the last three evaluations of χ^2.

When the minimum in χ^2 along the gradient is straddled, as denoted by an increase in χ^2, the parabolic interpolation of Equation (11-4) is computed in statements 91–93. The fitting function YFIT(I) = $y(x_I)$ and the value of CHISQR = χ_ν^2 at the final point are evaluated for return to the calling program. The main program is responsible for examining the change in χ^2 from a previous iteration to decide whether another iteration along a new gradient is advisable. For example, if the new value of χ^2 differs by less than 1% from the previous value, the search should be terminated.

Program 11-3   GRADLS   *(continued)*

```
      SUBROUTINE GRADLS (X, Y, SIGMAY, NPTS, NTERMS, MODE, A, DELTAA,
     1 YFIT, CHISQR)
      DIMENSION X(1), Y(1), SIGMAY(1), A(1), DELTAA(1), YFIT(1)
      DIMENSION GRAD(10)
C
C         EVALUATE CHI SQUARE AT BEGINNING
C
   11 NFREE = NPTS - NTERMS
      IF (NFREE) 13, 13, 21
   13 CHISQR = 0.
      GO TO 110
   21 DO 22 I=1, NPTS
   22 YFIT(I) = FUNCTN (X, I, A)
      CHISQ1 = FCHISQ (Y, SIGMAY, NPTS, NFREE, MODE, YFIT)
C
C         EVALUATE GRADIENT OF CHI SQUARE
C
   31 SUM = 0.
   32 DO 39 J=1, NTERMS
      DELTA = 0.1 * DELTAA(J)
      A(J) = A(J) + DELTA
      DO 36 I=1, NPTS
   36 YFIT(I) = FUNCTN (X, I, A)
      A(J) = A(J) - DELTA
      GRAD(J) = CHISQ1 - FCHISQ (Y, SIGMAY, NPTS, NFREE, MODE, YFIT)
   39 SUM = SUM + GRAD(J)**2
   41 DO 42 J=1, NTERMS
   42 GRAD(J) = DELTAA(J) * GRAD(J)/SQRT (SUM)
C
C         EVALUATE CHI SQUARE AT NEW POINT
C
   51 DO 52 J=1, NTERMS
   52 A(J) = A(J) + GRAD(J)
   53 DO 54 I=1, NPTS
   54 YFIT(I) = FUNCTN (X, I, A)
      CHISQ2 = FCHISQ (Y, SIGMAY, NPTS, NFREE, MODE, YFIT)
C
C         MAKE SURE CHI SQUARE DECREASES
C
   61 IF (CHISQ1 - CHISQ2) 62, 62, 71
   62 DO 64 J=1, NTERMS
      A(J) = A(J) - GRAD(J)
   64 GRAD(J) = GRAD(J) / 2.
      GO TO 51
C
C         INCREMENT PARAMETERS UNTIL CHI SQUARE STARTS TO INCREASE
C
   71 DO 72 J=1, NTERMS
   72 A(J) = A(J) + GRAD(J)
      DO 74 I=1, NPTS
   74 YFIT(I) = FUNCTN (X, I, A)
   75 CHISQ3 = FCHISQ (Y, SIGMAY, NPTS, NFREE, MODE, YFIT)
   76 IF (CHISQ3 - CHISQ2) 81, 91, 91
   81 CHISQ1 = CHISQ2
   82 CHISQ2 = CHISQ3
      GO TO 71
```

Program 11-3   GRADLS   *(continued)*

```
C
C           FIND MINIMUM OF PARABOLA DEFINED BY LAST THREE POINTS
C
   91 DELTA = 1./(1.+(CHISQ1-CHISQ2)/(CHISQ3-CHISQ2)) + 0.5
      DO 93 J=1, NTERMS
   93 A(J) = A(J) - DELTA*GRAD(J)
      DO 95 I=1, NPTS
   95 YFIT(I) = FUNCTN (X, I, A)
      CHISQR = FCHISQ (Y, SIGMAY, NPTS, NFREE, MODE, YFIT)
  101 IF (CHISQ2 - CHISQR) 102, 110, 110
  102 DO 103 J=1, NTERMS
  103 A(J) = A(J) + (DELTA-1.)*GRAD(J)
  104 DO 105 I=1, NPTS
  105 YFIT(I) = FUNCTN (X, I, A)
  106 CHISQR = CHISQ2
  110 RETURN
      END
```

Sample calculation   Table 11-1 gives the results of a gradient-search least-squares fitting applied to the data of Example 11-1. The starting values for the parameters a_j and the step sizes Δa_j are the same as for the grid search. The values listed at the end of each iteration correspond to the values at the end of successive passes through Program 11-2 at the minima of the parabolic interpolations. Note that the gradient search locates the bottom of the minimum more quickly than does the grid search, although the grid search gets fairly close with much less computation.

11-3   PARABOLIC EXTRAPOLATION OF x^2

Instead of searching the χ^2 hypersurface to map the variation of χ^2 with parameters, we should be able to find an approximate analytical function which describes the χ^2 hypersurface and to use this function to locate the minimum directly. Discrepancies in the approximations would yield errors in the calculated values of the parameters, but successive applications of the analytical method should approach the minimum of χ^2 with increasing accuracy.

The main advantage in such an approach is that the number of points on the χ^2 hypersurface at which computations must be made will be fewer than with a grid or gradient search. This advantage is somewhat compensated for by the fact that the computations at each point are considerably more complicated.

However, the analytical solution essentially chooses its own step size, and, thus, the user is spared the problem of trying to optimize the step size for speed and precision.

In this section we will describe a method of expanding the function χ^2 using an analytical expression for the variation of χ^2 to locate the minimum directly. In the next section we will discuss the alternative approach of expanding the fitting function $y(x)$ as a function of the parameters a_j in order to use the method of linear least-squares fitting.

Expansion of χ^2 Let us expand the function χ^2 to first order in a Taylor's series expansion as a function of the parameters a_j

$$\chi^2 = \chi_0^2 + \sum_{j=1}^{n} \left(\frac{\partial \chi_0^2}{\partial a_j} \delta a_j \right) \tag{11-9}$$

where χ_0^2 is the value of χ^2 at some starting point where the fitting function is $y_0(x)$

$$\chi_0^2 = \sum \left\{ \frac{1}{\sigma_i{}^2} [y_i - y_0(x_i)]^2 \right\} \tag{11-10}$$

and the δa_j are increments in the parameters a_j to reach the point at which $y(x)$ and χ^2 are to be evaluated.

Using the method of least squares, the optimum values for the parameter increments δa_j are those for which the function χ^2 is at a minimum in parameter space, i.e., for which the derivatives with respect to the parameters are 0.

$$\frac{\partial \chi^2}{\partial a_k} = \frac{\partial \chi_0^2}{\partial a_k} + \sum_{j=1}^{n} \left(\frac{\partial^2 \chi_0^2}{\partial a_j \partial a_k} \delta a_j \right) = 0 \qquad k = 1, n \tag{11-11}$$

The result is a set of n simultaneous linear equations in δa_j which we can treat, as in Chapters 8 and 9, as a matrix equation,

$$\beta_k = \sum_{j=1}^{n} (\delta a_j \, \alpha_{jk}) \qquad k = 1, n$$
$$\beta = \delta a \, \alpha \tag{11-12}$$

where β is a row matrix whose elements are equal (except for sign) to half the first term in Equations (11-11), and α is a symmetric matrix of order n whose elements are equal to half the coefficients of δa_j in Equations (11-11).

$$\beta_k \equiv -\frac{1}{2}\frac{\partial \chi_0^2}{\partial a_k} \qquad \alpha_{jk} \equiv \frac{1}{2}\frac{\partial^2 \chi_0^2}{\partial a_j \, \partial a_k} \qquad (11\text{-}13)$$

Parabolic expansion The solution of Equations (11-12) is equivalent to approximating the χ^2 hypersurface with a parabolic surface, even though the derivation uses only a first-order expansion of χ^2. To show this equivalence, let us expand the function χ^2 to second order in a Taylor's expansion as a function of the parameters a_j.

$$\chi^2 = \chi_0^2 + \sum_{j=1}^{n}\left(\frac{\partial \chi_0^2}{\partial a_j}\,\delta a_j\right) + \frac{1}{2}\sum_{j=1}^{n}\sum_{k=1}^{n}\left(\frac{\partial^2 \chi_0^2}{\partial a_j \, \partial a_k}\,\delta a_j\,\delta a_k\right) \quad (11\text{-}14)$$

The result is a function which is second order in the parameter increments δa_j and therefore describes a parabolic hypersurface.

The optimum values for the increments δa_j are those for which χ^2 is a minimum. We can obtain a solution by requiring that the derivatives with respect to the *increments* be equal to 0.

$$\frac{\partial \chi^2}{\partial \delta a_k} = \frac{\partial \chi_0^2}{\partial a_k} + \sum_{j=1}^{n}\left(\frac{\partial^2 \chi_0^2}{\partial a_j \, \partial a_k}\,\delta a_j\right) = 0 \qquad k = 1, n \qquad (11\text{-}15)$$

Comparison of Equations (11-11) and (11-15) shows that the two methods yield identical results. Setting the derivatives of χ^2 of Equation (11-14) with respect to the *parameters* equal to 0 would yield the same result if higher-order terms in δa_j are neglected.

Error matrix Let us compare the solutions of Equations (11-12) with the analogous solutions for linear least-squares fitting of Equations (8-22). A comparison of the definitions of α_{jk} in Equations (11-13) and (8-24) shows that the two are identical, and the symmetric matrix α is therefore the *curvature matrix* discussed in Section 8-3, so named because it measures the curvature of the χ^2 hypersurface.

To compare the definitions of β_k, let us take the derivative of χ_0^2 in Equation (11-10).

$$\beta_k \equiv -\frac{1}{2}\frac{\partial \chi_0^2}{\partial a_k} = \sum \left\{ \frac{1}{\sigma_i^2}[y_i - y_0(x_i)]\frac{\partial y_0(x_i)}{\partial a_k} \right\} \tag{11-16}$$

This is equivalent to the definition of β_k in Equations (8-23) except for the substitution of $y_i - y_0(x_i)$ for y_i. We can easily justify this substitution by noting that the solution of Equations (11-12) is for the parameter increments δa_j, whereas that of Equations (8-22) is for the parameters themselves. In essence, we are using linear least-squares methods to fit the parameter increments to difference data y_i' between the actual data and the starting values of the fitting function $y_0(x_i)$.

$$y_i' = y_i - y_0(x_i) \tag{11-17}$$

With this modification, the results are identical.

Methods of computation   The solution to matrix (11-12) can be obtained by matrix inversion as in Chapters 8 and 9

$$\delta a = \beta \epsilon \qquad \delta a_j = \sum_{k=1}^{n}(\beta_k \epsilon_{jk}) \tag{11-18}$$

where the error matrix $\epsilon = \alpha^{-1}$ is the inverse of the curvature matrix.

The method of determinants for solving simultaneous equations discussed in Appendix B and Chapters 6 and 8 can also be applied to Equations (11-12). The solution for δa_1 for $n = 3$, for example, is

$$\delta a_1 = \frac{1}{\Delta}\begin{vmatrix} \beta_1 & \alpha_{12} & \alpha_{13} \\ \beta_2 & \alpha_{22} & \alpha_{23} \\ \beta_3 & \alpha_{32} & \alpha_{33} \end{vmatrix} \qquad \Delta = \begin{vmatrix} \alpha_{11} & \alpha_{12} & \alpha_{13} \\ \alpha_{21} & \alpha_{22} & \alpha_{23} \\ \alpha_{31} & \alpha_{32} & \alpha_{33} \end{vmatrix}$$

or, in general,

$$\delta a_j = \frac{|\alpha'|}{|\alpha|} \qquad \text{where} \begin{cases} \alpha_{km}' = \alpha_{km} & \text{for } m \neq j \\ \alpha_{kj}' = \beta_k \end{cases} \tag{11-19}$$

If the parameters a_j are independent of each other, i.e., if the variation of χ^2 with respect to each parameter is independent of the values of the other parameters, then the cross partial derivatives $\alpha_{jk}(j \neq k)$ are 0 and the denominator of Equation (11-19) is a diagonal determinant. The solutions become considerably simplified because the matrix equation of Equations (11-12) degenerates into n separate equations.

$$\delta a_j \simeq \frac{\beta_j}{\alpha_{jj}} = -\frac{\partial \chi_0^2 / \partial a_j}{\partial^2 \chi_0^2 / \partial a_j{}^2} \tag{11-20}$$

Computation of the elements of the matrices of Equations (11-13) may be approximated by determining the variation of χ^2 in the neighborhood of the starting point χ_0^2.

$$\frac{\partial \chi_0^2}{\partial a_j} \simeq \frac{\chi_0^2(a_j + \Delta a_j, a_k) - \chi_0^2(a_j - \Delta a_j, a_k)}{2\Delta a_j}$$

$$\frac{\partial^2 \chi_0^2}{\partial a_j{}^2} \simeq \frac{\chi_0^2(a_j + \Delta a_j, a_k) - 2\chi_0^2(a_j, a_k) + \chi_0^2(a_j - \Delta a_j, a_k)}{\Delta a_j{}^2}$$

$$\frac{\partial^2 \chi_0^2}{\partial a_j\, \partial a_k} \simeq \frac{\begin{array}{c}\chi_0^2(a_j + \Delta a_j, a_k + \Delta a_k) - \chi_0^2(a_j + \Delta a_j, a_k) \\ - \chi_0^2(a_j, a_k + \Delta a_k) + \chi_0^2(a_j, a_k)\end{array}}{\Delta a_j\, \Delta a_k}$$

$$\tag{11-21}$$

The Δa_j are step sizes which must be large enough to prevent round-off error in the computation and small enough to furnish reasonable answers near the minimum where the derivatives may be changing rapidly with parameters.

Note the symmetry of the variation of the parameters around the starting point for the computation of the first derivatives, but not for the second derivatives which are assumed to be changing slowly in this approximation. The points at which χ^2 must be computed are indicated in Figure 11-6. The solid dots are the points used for the determination of the first derivatives; the open dots are used in addition for determination of the second derivatives.

Search procedure Within the limits of the approximation of the χ^2 hypersurface with a parabolic extrapolation, we can

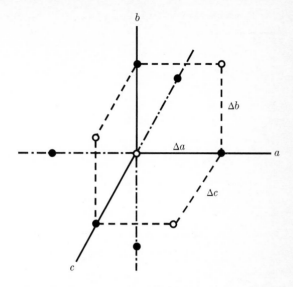

FIGURE 11-6 Values of parameters a, b, and c at which χ^2 must be evaluated for computation of derivatives of χ^2. First derivatives are evaluated using solid points only; second derivatives require values of χ^2 at all points.

solve Equations (11-12) directly to yield parameter increments δa_j such that χ^2 should be minimized for $a_j + \delta a_j$. If the starting point is close enough to the minimum that higher-order terms in the expansion can be neglected, this becomes an accurate and precise method. But if the starting point is not near enough, the parabolic approximation of the χ^2 hypersurface is not valid and the results will be in error, generally in the direction that the parameter increment δa_j is too large.

In fact, if the starting point is so far from the minimum that the curvature of χ^2 is negative, the solution will tend to increment the parameter with the wrong sign forcing χ^2 towards a maximum rather than a minimum. During computation, therefore, all second partial derivatives with respect to one parameter α_{jj} must be considered as positive whether they are or not. The resulting

Program 11-4   CHIFIT   Least-squares fit by parabolic extrapolation of χ^2.

```
C      SUBROUTINE CHIFIT
C
C      PURPOSE
C        MAKE A LEAST-SQUARES FIT TO A NON-LINEAR FUNCTION
C          WITH A PARABOLIC EXPANSION OF CHI SQUARE
C
C      USAGE
C        CALL CHIFIT (X, Y, SIGMAY, NPTS, NTERMS, MODE, A, DELTAA,
C          SIGMAA, YFIT, CHISQR)
C
C      DESCRIPTION OF PARAMETERS
C        X      - ARRAY OF DATA POINTS FOR INDEPENDENT VARIABLE
C        Y      - ARRAY OF DATA POINTS FOR DEPENDENT VARIABLE
C        SIGMAY - ARRAY OF STANDARD DEVIATIONS FOR Y DATA POINTS
C        NPTS   - NUMBER OF PAIRS OF DATA POINTS
C        NTERMS - NUMBER OF PARAMETERS
C        MODE   - DETERMINES METHOD OF WEIGHTING LEAST-SQUARES FIT
C                 +1 (INSTRUMENTAL) WEIGHT(I) = 1./SIGMAY(I)**2
C                  0 (NO WEIGHTING)  WEIGHT(I) = 1.
C                 -1 (STATISTICAL)  WEIGHT(I) = 1./Y(I)
C        A      - ARRAY OF PARAMETERS
C        DELTAA - ARRAY OF INCREMENTS FOR PARAMETERS A
C        SIGMAA - ARRAY OF STANDARD DEVIATIONS FOR PARAMETERS A
C        YFIT   - ARRAY OF CALCULATED VALUES OF Y
C        CHISQR - REDUCED CHI SQUARE FOR FIT
C
C      SUBROUTINES AND FUNCTION SUBPROGRAMS REQUIRED
C        FUNCTN (X, I, A)
C          EVALUATES THE FITTING FUNCTION FOR THE ITH TERM
C        FCHISQ (Y, SIGMAY, NPTS, NFREE, MODE, YFIT)
C          EVALUATES REDUCED CHI SQUARE FOR FIT TO DATA
C        MATINV (ARRAY, NTERMS, DET)
C          INVERTS A SYMMETRIC TWO-DIMENSIONAL MATRIX OF DEGREE NTERMS
C          AND CALCULATES ITS DETERMINANT
C
C      MODIFICATIONS FOR FORTRAN II
C        OMIT DOUBLE PRECISION SPECIFICATIONS
C        ADD F SUFFIX TO SQRT IN STATEMENT 104
C
C      COMMENTS
C        DIMENSION STATEMENT VALID FOR NTERMS UP TO 10
C
```

magnitude for δa_j will be incorrect, but the sign will be correct.

Since the result tends to be too large, a reasonable perturbation on the search procedure is to divide the increment δa_j into fractional steps $f\ \delta a_j$, where f is on the order of 0.2, and search along that vector until the minimum is straddled. Parabolic interpolation can then be used to locate the bottom of the minimum.

A typical search path is indicated in Table 11-1 for the data of Example 11-1 for comparison with the methods of grid and

```
      SUBROUTINE CHIFIT (X, Y, SIGMAY, NPTS, NTERMS, MODE, A, DELTAA,
     1 SIGMAA, YFIT, CHISQR)
      DOUBLE PRECISION ALPHA
      DIMENSION X(1), Y(1), SIGMAY(1), A(1), DELTAA(1), SIGMAA(1),
     1 YFIT(1)
      DIMENSION ALPHA(10,10), BETA(10), DA(10)
   11 NFREE = NPTS - NTERMS
      FREE = NFREE
      IF (NFREE) 14, 14, 16
   14 CHISQR = 0.
      GO TO 120
   16 DO 17 I=1, NPTS
   17 YFIT(I) = FUNCTN (X, I, A)
      CHISQ1 = FCHISQ (Y, SIGMAY, NPTS, NFREE, MODE, YFIT)
C
C        EVALUATE ALPHA AND BETA MATRICES
C
   20 DO 60 J=1, NTERMS
C
C           A(J) + DELTAA(J)
C
   21 AJ = A(J)
      A(J) = AJ + DELTAA(J)
      DO 24 I=1, NPTS
   24 YFIT(I) = FUNCTN (X, I, A)
      CHISQ2 = FCHISQ (Y, SIGMAY, NPTS, NFREE, MODE, YFIT)
      ALPHA(J,J) = CHISQ2 - 2.*CHISQ1
      BETA(J) = -CHISQ2
   31 DO 50 K=1, NTERMS
      IF (K - J) 33, 50, 36
   33 ALPHA(K,J) = (ALPHA(K,J) - CHISQ2) / 2.
      ALPHA(J,K) = ALPHA(K,J)
      GO TO 50
   36 ALPHA(J,K) = CHISQ1 - CHISQ2
C
C           A(J) + DELTAA(J)  AND  A(K) + DELTAA(K)
C
   41 AK = A(K)
      A(K) = AK + DELTAA(K)
      DO 44 I=1, NPTS
   44 YFIT(I) = FUNCTN (X, I, A)
      CHISQ3 = FCHISQ (Y, SIGMAY, NPTS, NFREE, MODE, YFIT)
      ALPHA(J,K) = ALPHA(J,K) + CHISQ3
      A(K) = AK
   50 CONTINUE
C
C           A(J) - DELTAA(J)
C
   51 A(J) = AJ - DELTAA(J)
      DO 53 I=1, NPTS
   53 YFIT(I) = FUNCTN (X, I, A)
      CHISQ3 = FCHISQ (Y, SIGMAY, NPTS, NFREE, MODE, YFIT)
      A(J) = AJ
      ALPHA(J,J) = (ALPHA(J,J) + CHISQ3) / 2.
      BETA(J) = (BETA(J) + CHISQ3) / 4.
   60 CONTINUE
```

gradient search. Note that the start of the search is even more erratic than any of the other methods, but the minimum of χ^2 is located quite precisely once the search converges.

```
C
C          ELIMINATE NEGATIVE CURVATURE
C
   61 DO 70 J=1, NTERMS
      IF (ALPHA(J,J)) 63, 65, 70
   63 ALPHA(J,J) = -ALPHA(J,J)
      GO TO 66
   65 ALPHA(J,J) = 0.01
   66 DO 70 K=1, NTERMS
      IF (K - J) 68, 70, 68
   68 ALPHA(J,K) = 0.
      ALPHA(K,J) = 0.
   70 CONTINUE
C
C          INVERT MATRIX AND EVALUATE PARAMETER INCREMENTS
C
   71 CALL MATINV (ALPHA, NTERMS, DET)
      DO 76 J=1, NTERMS
      DA(J) = 0.
   74 DO 75 K=1, NTERMS
   75 DA(J) = DA(J) + BETA(K)*ALPHA(J,K)
   76 DA(J) = 0.2 * DA(J) * DELTAA(J)
C
C          MAKE SURE CHI SQUARE DECREASES
C
   81 DO 82 J=1, NTERMS
   82 A(J) = A(J) + DA(J)
   83 DO 84 I=1, NPTS
   84 YFIT(I) = FUNCTN (X, I, A)
      CHISQ2 = FCHISQ (Y, SIGMAY, NPTS, NFREE, MODE, YFIT)
      IF (CHISQ1 - CHISQ2) 87, 91, 91
   87 DO 89 J=1, NTERMS
      DA(J) = DA(J)/2.
   89 A(J) = A(J) - DA(J)
      GO TO 83
C
C          INCREMENT PARAMETERS UNTIL CHI SQUARE STARTS TO INCREASE
C
   91 DO 92 J=1, NTERMS
   92 A(J) = A(J) + DA(J)
      DO 94 I=1, NPTS
   94 YFIT(I) = FUNCTN (X, I, A)
      CHISQ3 = FCHISQ (Y, SIGMAY, NPTS, NFREE, MODE, YFIT)
      IF (CHISQ3 - CHISQ2) 97, 101, 101
   97 CHISQ1 = CHISQ2
      CHISQ2 = CHISQ3
   99 GO TO 91
```

Program 11-4   The method of least-squares fitting discussed in this section is illustrated in the computer routine CHIFIT of Program 11-4. This is a Fortran subroutine which uses the method of parabolic extrapolation of χ^2 to make a least-squares fit to a function which need not be linear in its parameters. The input variables are X, Y, SIGMAY, NPTS, NTERMS, MODE, A, and DELTAA, and the output variables are A, SIGMAA, YFIT, and CHISQR.

```
C
C           FIND MINIMUM OF PARABOLA DEFINED BY LAST THREE POINTS
C
  101 DELTA = 1./(1.+(CHISQ1-CHISQ2)/(CHISQ3-CHISQ2)) + 0.5
      DO 104 J=1, NTERMS
      A(J) = A(J) - DELTA*DA(J)
  104 SIGMAA(J) = DELTAA(J) * SQRT (FREE*ALPHA(J,J))
      DO 106 I=1, NPTS
  106 YFIT(I) = FUNCTN (X, I, A)
      CHISQR = FCHISQ (Y, SIGMAY, NPTS, NFREE, MODE, YFIT)
  111 IF (CHISQ2 - CHISQR) 112, 120, 120
  112 DO 113 J=1, NTERMS
  113 A(J) = A(J) + (DELTA-1.)*DA(J)
      DO 115 I=1, NPTS
  115 YFIT(I) = FUNCTN (X, I, A)
      CHISQR = CHISQ2
  120 RETURN
      END
```

See the discussion of Program 11-1 for a description of these variables and of the function FUNCTN which must be supplied in addition.

First and second derivatives of χ^2 are calculated in the DO loop of statements 20–60 according to Equations (11-21) and deposited in the arrays ALPHA(J,K) = α_{JK} and BETA(J) = β_J according to Equations (11-13). The values of χ^2 are evaluated with the routine FCHISQ of Program 10-2. The subroutine MATINV of Appendix B is used to invert the matrix α in statement 71, and the parameter increments DA(J) = δa_J are accumulated in statements 74–76 according to Equation (11-18). Note that the elements of the matrices are multiplied by DELTAA(J) DELTAA(K) to improve the precision of the matrix inversion.

Instead of using the increments δa_j directly, the program searches along the same direction in steps of $\delta a_j/5$ (from statement 76) in statements 91–99, monitoring the variation of χ^2 until χ^2 starts to increase. The size of the increments is decreased still further in statements 87–89 if the first increment in a_j does not yield a decrease in χ^2. The last three points of the search are used in statements 101–104 to locate the point of minimum χ^2 by parabolic interpolation, provided uncertainties in the interpolation do not yield a higher value of χ^2 than the lowest value already found.

11-4 LINEARIZATION OF FUNCTION

An alternative to expanding the function χ^2 to develop an analytical description of the hypersurface is to expand the fitting function $y(x)$ as a function of the parameters a_j and use the method of linear least squares to determine the optimum value for the parameter increments δa_j. If we carry out the derivation rigorously and drop high-order terms, we should achieve the same result as in Section 11-3 as the expansion of χ^2 to first and second order did.

First-order expansion   Let us expand the fitting function $y(x)$ to first order in a Taylor's expansion as a function of the parameters a_j.

$$y(x) = y_0(x) + \sum_{j=1}^{n} \left[\frac{\partial y_0(x)}{\partial a_j} \delta a_j \right]$$

The result is a function which is linear in the parameter increments δa_j to which we can apply the method of linear least squares developed in Chapters 8 and 9.

The derivatives are assumed to be evaluated at the starting point $y_0(x)$. For computational purposes, they may be calculated either by using the analytical expressions for the fitting function $y(x)$ or from an empirical determination of the variation of $y(x)$ with the parameters δa_j as in Equation (11-5).

$$\frac{\partial y_0(a_j)}{\partial a_j} \sim \frac{y_0(a_j + \Delta a_j) - y_0(a_j - \Delta a_j)}{2\Delta a_j} \tag{11-22}$$

To this approximation, χ^2 can be expressed explicitly as a function of the parameter increments δa_j.

$$\chi^2 = \sum \left(\frac{1}{\sigma_i^2} \left\{ y_i - y_0(x_i) - \sum_{j=1}^{n} \left[\frac{\partial y_0(x_i)}{\partial a_j} \delta a_j \right] \right\}^2 \right) \tag{11-23}$$

One way of describing this approach is to note that if we define a new set of difference data y_i' as in Equation (11-17), we can fit these data with a linear difference function $y'(x)$

$$y'(x) = \sum_{j=1}^{n} \left[\frac{\partial y_0(x_i)}{\partial a_j} \delta a_j \right]$$

with coefficients $a_j' = \delta a_j$ and fitting functions

$$X_j(x_i) = \partial y_0(x_i)/\partial a_j$$

Following the method of linear least squares, we minimize χ^2 with respect to each of the parameter increments δa_j by setting the derivatives equal to 0.

$$\frac{\partial \chi^2}{\partial \delta a_k} = -2 \sum \left(\frac{1}{\sigma_i^2} \left\{ y_i - y_0(x_i) - \sum_{j=1}^{n} \left[\frac{\partial y_0(x_i)}{\partial a_j} \delta a_j \right] \right\} \frac{\partial y_0(x_i)}{\partial a_k} \right)$$
$$= 0$$

As before, this yields a set of n simultaneous equations

$$\beta_k = \sum_{j=1}^{n} (\delta a_j \, \alpha_{jk}) \qquad k = 1, \, n \tag{11-24}$$
$$\beta = \delta a \, \alpha$$

where β_k is defined as in Equation (11-16) and α_{jk} is given by

$$\alpha_{jk} \simeq \sum \left[\frac{1}{\sigma_i^2} \frac{\partial y_0(x_i)}{\partial a_j} \frac{\partial y_0(x_i)}{\partial a_k} \right] \tag{11-25}$$

A comparison with the definition of α_{jk} in Equation (8-23) shows that the two are identical, at least in the linear approximation.

Second-order expansion Suppose we expand the fitting function $y(x)$ to second order in a Taylor's expansion as a function of the parameters a_j.

$$y(x) = y_0(x) + \sum_{j=1}^{n} \left[\frac{\partial y_0(x)}{\partial a_j} \delta a_j \right] + \frac{1}{2} \sum_{j=1}^{n} \sum_{k=1}^{n} \left[\frac{\partial^2 y_0(x)}{\partial a_j \, \partial a_k} \delta a_j \, \delta a_k \right]$$

If we set the derivatives of χ^2 with respect to the parameter increments δa_j equal to 0 and neglect all terms of higher than linear in

δa_j, the result is the same as if we take the derivative of Equation (11-23) with respect to the parameters a_j.

$$\beta_k = \sum_{j=1}^{n} (\delta a_j \, \alpha_{jk}) \qquad k = 1, n$$

$$\beta_k \equiv \sum \left\{ \frac{1}{\sigma_i{}^2} [y_i - y_0(x_i)] \frac{\partial y_0(x_i)}{\partial a_k} \right\} = -\frac{1}{2} \frac{\partial \chi_0^2}{\partial a_k}$$

$$\alpha_{jk} \equiv \sum \left(\frac{1}{\sigma_i{}^2} \left\{ \frac{\partial y_0(x_i)}{\partial a_j} \frac{\partial y_0(x_i)}{\partial a_k} - [y_i - y_0(x_i)] \frac{\partial^2 y_0(x_i)}{\partial a_j \, \partial a_k} \right\} \right)$$

$$= \frac{1}{2} \frac{\partial^2 \chi_0^2}{\partial a_j \, \partial a_k}$$

(11-26)

These definitions are identical to those in Equations (11-13).

Thus, the solution given in Equation (11-25) for α_{jk} is a first-order approximation to the curvature matrix which is given correctly in Equations (11-26). For linear functions the second-order term vanishes; hence the ambiguity in whether the definition of Equation (11-25) is identical to that of Equation (8-23).

Convergence One disadvantage inherent in the analytical methods of expanding either the fitting function $y(x)$ or χ^2 is that, while they converge quite rapidly to the point of minimum χ^2 from points nearby, they cannot be relied on to approach the minimum with any accuracy from a point outside the region where the χ^2 hypersurface is approximately parabolic. In particular, if the curvature of the χ^2 hypersurface is used, as in Equations (11-13) or (11-26), the analytical solution is unreliable whenever the curvature becomes negative. Symptomatic of this problem are the two checks built into the program CHIFIT: one to treat all curvatures as if they were positive, and the other to search along the path of the analytical solution and demand that χ^2 decrease at the first step.

In contrast, the gradient search of Section 11-2 is ideally suited for approaching the minimum from far away, but it does not converge rapidly when in the immediate neighborhood. Therefore, we need an algorithm which behaves like a gradient search for the first portion of a search and behaves more like an analytical solution as the search converges. In fact, it can be shown (see Marquardt) that the path directions for gradient and analytical searches are nearly perpendicular to each other, and that the optimum direction is somewhere between these two vectors.

One advantage of combining these two methods into one algorithm is that the simpler first-order expansion of the analytical method will suffice since the expansion need only be valid in the immediate neighborhood of the minimum. Thus, we can neglect the second derivatives of Equations (11-26) and use the approximation of Equation (11-25) for a determination of the curvature matrix α.

Gradient-expansion algorithm A convenient algorithm (see Marquardt) which combines the best features of the gradient search with the method of linearizing the fitting function can be obtained by increasing the diagonal terms of the curvature matrix α by a factor λ which controls the interpolation of the algorithm between the two extremes. The matrix equation of Equation (11-24) becomes

$$
\beta = \delta a \, \alpha'
$$
$$
\alpha'_{jk} = \begin{cases} \alpha_{jk}(1 + \lambda) & \text{for } j = k \\ \alpha_{jk} & \text{for } j \neq k \end{cases} \tag{11-27}
$$

If λ is very small, Equations (11-27) are similar to the solution of Equations (11-24) developed from a Taylor's expansion. If λ is very large, the diagonal terms of the curvature matrix dominate and the matrix equation degenerates into n separate equations

$$
\beta_j \simeq \lambda \, \delta a_j \, \alpha_{jj} \tag{11-28}
$$

which yield increments δa_j in the same direction as the gradients β_j of Equation (11-16) but with lengths scaled by α_{jj} and reduced by a factor of λ.

If the full curvature matrix of Equations (11-26) is used for α_{jj}, Equation (11-28) is identical to Equation (11-20) and the algorithm resembles the analytical solution of expanding χ^2, ignoring the interdependence of parameters. If we use the approximation of Equation (11-25), however, the factor of α_{jj} in Equation (11-28) is simply a scale factor for the gradient equal to the standard deviation of the derivatives $\partial y(x_i)/\partial a_j$. Therefore, we shall assume this approximation, which also obviates the use of second derivatives.

The solution for the parameter increments δa_j follows from Equations (11-27) after matrix inversion

$$\delta a_j = \sum_{k=1}^{n} (\beta_k \epsilon'_{jk}) \tag{11-29}$$

where the β_k are given by Equation (11-16) and the matrix ϵ' is the inverse of the matrix α' whose elements are given in Equations (11-27).

The value of the constant factor λ should be chosen small enough to take advantage of the analytical solution, but large enough that χ^2 decreases. Since this algorithm approaches the gradient-search method with small steps for large λ, there should exist a value of λ such that $\chi^2(a + \delta a) < \chi^2(a)$. The recipe given by Marquardt is:

1. Compute $\chi^2(a)$.
2. Start initially with $\lambda = 0.001$.
3. Compute δa and $\chi^2(a + \delta a)$ with this choice of λ.
4. If $\chi^2(a + \delta a) > \chi^2(a)$, increase λ by a factor of 10 and repeat step (3).
5. If $\chi^2(a + \delta a) < \chi^2(a)$, decrease λ by a factor of 10, consider $a' = a + \delta a$ to be the new starting point, and return to step (3) substituting a' for a.

For each iteration it may be necessary to recompute the parameter increments δa_j from Equation (11-29) several times to optimize λ. The accumulation of the elements of the matrices α_{jk} and β_j need be done only once per iteration, however. As the solution approaches the minimum, the value of λ will decrease and the solution will locate the minimum with few iterations. A lower

Program 11-5   CURFIT Least-squares fit by linearization of fitting function.

```
C     SUBROUTINE CURFIT
C
C     PURPOSE
C        MAKE A LEAST-SQUARES FIT TO A NON-LINEAR FUNCTION
C           WITH A LINEARIZATION OF THE FITTING FUNCTION
C
C     USAGE
C        CALL CURFIT (X, Y, SIGMAY, NPTS, NTERMS, MODE, A, DELTAA,
C           SIGMAA, FLAMDA, YFIT, CHISQR)
C
C     DESCRIPTION OF PARAMETERS
C        X      - ARRAY OF DATA POINTS FOR INDEPENDENT VARIABLE
C        Y      - ARRAY OF DATA POINTS FOR DEPENDENT VARIABLE
C        SIGMAY - ARRAY OF STANDARD DEVIATIONS FOR Y DATA POINTS
C        NPTS   - NUMBER OF PAIRS OF DATA POINTS
C        NTERMS - NUMBER OF PARAMETERS
C        MODE   - DETERMINES METHOD OF WEIGHTING LEAST-SQUARES FIT
C                 +1 (INSTRUMENTAL) WEIGHT(I) = 1./SIGMAY(I)**2
C                  0 (NO WEIGHTING) WEIGHT(I) = 1.
C                 -1 (STATISTICAL)  WEIGHT(I) = 1./Y(I)
C        A      - ARRAY OF PARAMETERS
C        DELTAA - ARRAY OF INCREMENTS FOR PARAMETERS A
C        SIGMAA - ARRAY OF STANDARD DEVIATIONS FOR PARAMETERS A
C        FLAMDA - PROPORTION OF GRADIENT SEARCH INCLUDED
C        YFIT   - ARRAY OF CALCULATED VALUES OF Y
C        CHISQR - REDUCED CHI SQUARE FOR FIT
C
C     SUBROUTINES AND FUNCTION SUBPROGRAMS REQUIRED
C        FUNCTN (X, I, A)
C           EVALUATES THE FITTING FUNCTION FOR THE ITH TERM
C        FCHISQ (Y, SIGMAY, NPTS, NFREE, MODE, YFIT)
C           EVALUATES REDUCED CHI SQUARE FOR FIT TO DATA
C        FDERIV (X, I, A, DELTAA, NTERMS, DERIV)
C           EVALUATES THE DERIVATIVES OF THE FITTING FUNCTION
C           FOR THE ITH TERM WITH RESPECT TO EACH PARAMETER
C        MATINV (ARRAY, NTERMS, DET)
C           INVERTS A SYMMETRIC TWO-DIMENSIONAL MATRIX OF DEGREE NTERMS
C           AND CALCULATES ITS DETERMINANT
C
C     MODIFICATIONS FOR FORTRAN II
C        OMIT DOUBLE PRECISION SPECIFICATIONS
C        ADD F SUFFIX TO SQRT IN STATEMENTS 73, 84, AND 103
C
C     COMMENTS
C        DIMENSION STATEMENT VALID FOR NTERMS UP TO 10
C        SET FLAMDA = 0.001 AT BEGINNING OF SEARCH
```

limit may be set for the value of λ, but in practice this limit will seldom be reached.

Program 11-5   The discussion of this section is illustrated in the computer routine CURFIT of Program 11-5. This is a Fortran subroutine to make a least-squares fit to a function, which need not be linear in its parameters, using the algorithm of Marquardt which combines a gradient search with an analytical solution

```
      SUBROUTINE CURFIT (X, Y, SIGMAY, NPTS, NTERMS, MODE, A, DELTAA,
     1 SIGMAA, FLAMDA, YFIT, CHISQR)
      DOUBLE PRECISION ARRAY
      DIMENSION X(1), Y(1), SIGMAY(1), A(1), DELTAA(1), SIGMAA(1),
     1 YFIT(1)
      DIMENSION WEIGHT(100), ALPHA(10,10), BETA(10), DERIV(10),
     1 ARRAY(10,10), B(10)
   11 NFREE = NPTS - NTERMS
      IF (NFREE) 13, 13, 20
   13 CHISQR = 0.
      GO TO 110
C
C         EVALUATE WEIGHTS
C
   20 DO 30 I=1, NPTS
   21 IF (MODE) 22, 27, 29
   22 IF (Y(I)) 25, 27, 23
   23 WEIGHT(I) = 1. / Y(I)
      GO TO 30
   25 WEIGHT(I) = 1. / (-Y(I))
      GO TO 30
   27 WEIGHT(I) = 1.
      GO TO 30
   29 WEIGHT(I) = 1. / SIGMAY(I)**2
   30 CONTINUE
C
C         EVALUATE ALPHA AND BETA MATRICES
C
   31 DO 34 J=1, NTERMS
      BETA(J) = 0.
      DO 34 K=1, J
   34 ALPHA(J,K) = 0.
   41 DO 50 I=1, NPTS
      CALL FDERIV (X, I, A, DELTAA, NTERMS, DERIV)
      DO 46 J=1, NTERMS
      BETA(J) = BETA(J) + WEIGHT(I)*(Y(I)-FUNCTN(X,I,A))*DERIV(J)
      DO 46 K=1, J
   46 ALPHA(J,K) = ALPHA(J,K) + WEIGHT(I)*DERIV(J)*DERIV(K)
   50 CONTINUE
   51 DO 53 J=1, NTERMS
      DO 53 K=1, J
   53 ALPHA(K,J) = ALPHA(J,K)
C
C         EVALUATE CHI SQUARE AT STARTING POINT
C
   61 DO 62 I=1, NPTS
   62 YFIT(I) = FUNCTN (X, I, A)
   63 CHISQ1 = FCHISQ (Y, SIGMAY, NPTS, NFREE, MODE, YFIT)
C
C         INVERT MODIFIED CURVATURE MATRIX TO FIND NEW PARAMETERS
C
   71 DO 74 J=1, NTERMS
      DO 73 K=1, NTERMS
   73 ARRAY(J,K) = ALPHA(J,K) / SQRT (ALPHA(J,J) * ALPHA(K,K))
   74 ARRAY(J,J) = 1. + FLAMDA
   80 CALL MATINV (ARRAY, NTERMS, DET)
   81 DO 84 J=1, NTERMS
      B(J) = A(J)
      DO 84 K=1, NTERMS
   84 B(J) = B(J) + BETA(K)*ARRAY(J,K)/SQRT (ALPHA(J,J)*ALPHA(K,K))
```

Program 11-5   CURFIT   *(continued)*

```
C
C          IF CHI SQUARE INCREASED, INCREASE FLAMDA AND TRY AGAIN
C
   91 DO 92 I=1, NPTS
   92 YFIT(I) = FUNCTN (X, I, B)
   93 CHISQR = FCHISQ (Y, SIGMAY, NPTS, NFREE, MODE, YFIT)
      IF (CHISQ1 - CHISQR) 95, 101, 101
   95 FLAMDA = 10.*FLAMDA
      GO TO 71
C
C          EVALUATE PARAMETERS AND UNCERTAINTIES
C
  101 DO 103 J=1, NTERMS
      A(J) = B(J)
  103 SIGMAA(J) = SQRT (ARRAY(J,J) / ALPHA(J,J))
      FLAMDA = FLAMDA/10.
  110 RETURN
      END
```

developed from linearizing the fitting function. The input variables (x, y, SIGMAY, NPTS, NTERMS, MODE, FLAMDA, A, and DELTAA), and the output variables (A, SIGMAA, YFIT, and CHISQR) are described in Program 11-1. FLAMDA $= \lambda$ is the constant factor which should be set equal to 0.001 at the beginning of each fit.

The elements of the matrices α_{jk} and β_j are accumulated in statements 41–53 using the weighting factors evaluated in statements 20–30. The fitting function is evaluated with the function FUNCTN(X,I,A) $= y(x_I)$ described in Program 11-2 where the A(J) $= a_J$ are the parameters. The derivatives are evaluated with the routine FDERIV(X,I,A,DELTAA,J) $= \partial y(x_i)/\partial a_j$ which will be described in Programs 11-6 and 11-7.

The value of CHISQ1 $= \chi^2(a)$ at the starting point is evaluated in statements 61–63. The matrix ARRAY(J,K) $= \alpha'_{JK}$ is evaluated from the matrix ALPHA(J,K) $= \alpha_{JK}$ according to Equations (11-27) and inverted in statement 80 with the subroutine MATINV of Appendix B. Elements of the matrices are normalized to the diagonal elements of ALPHA to improve the precision of the matrix inversion. New parameters B(J) $= a_J + \delta a_J$ are accumulated in statements 81–84. If the new value of

$$\text{CHISQR} = \chi^2(a + \delta a)$$

evaluated in statements 91–93 is greater than CHISQ1 $= \chi^2(a)$, the value of FLAMDA $= \lambda$ is increased by a factor of 10 and the step is repeated.

If the new value of χ^2 is lower, the new parameters B(J) are deposited in the parameter array A(J) and the routine returns to the main program. It is the responsibility of the main program to monitor the variation of χ^2 and terminate the fit after a sufficient number of iterations.

This algorithm is also illustrated in the Fortran program "Least-Squares Estimation of Nonlinear Parameters," available as IBM Share Program EID-NLIN No. 3094.01.

Program 11-6 The calculation of derivatives for CURFIT is illustrated in the computer routine FDERIV of Program 11-6. This is a Fortran subroutine to evaluate the derivatives of the Gaussian plus quadratic polynomial function of Equation (11-1). The input variables are X, I, A, DELTAA, and NTERMS, and the output variable is DERIV.

The array X contains the data points for the independent variable, and the array A contains the values of the parameters A(J) = a_J. The variable I = i is the index of the data point, and the variable NTERMS = n is the number of parameters with respect to which the derivative is to be taken. The array DELTAA is ignored in this version.

The derivatives are evaluated analytically

$$\text{DERIV (J)} = \frac{\partial y(x_I)}{\partial a_J}$$

with six different expressions for the six possible derivatives. The values of the derivatives are returned as the array DERIV.

EXAMPLE The speed and precision of this type of search are indicated in Table 11-1 for the data of Example 11-1. Note that this method converges faster and more reliably than any other.

Program 11-7 An alternative approach for a general method of evaluating derivatives empirically is illustrated in the computer routine FDERIV of Program 11-7. This is a Fortran subroutine to evaluate the derivative of a general function which need not be specified. The value of the function is computed for

Program 11-6 FDERIV (Gaussian + quadratic) Derivative of fitting function.

```
C       SUBROUTINE FDERIV (GAUSSIAN + QUADRATIC)
C
C       PURPOSE
C         EVALUATE DERIVATIVES OF FUNCTION FOR LEAST-SQUARES SEARCH
C             WITH FORM OF A GAUSSIAN PEAK PLUS QUADRATIC POLYNOMIAL
C             FUNCTN(X,I,A) = A(1)*EXP(-Z**2/2) + A(4) + A(5)*X + A(6)*X**2
C             WHERE X = X(I) AND Z = (X - A(2))/A(3)
C
C       USAGE
C         CALL FDERIV (X, I, A, DELTAA, NTERMS, DERIV)
C
C       DESCRIPTION.OF PARAMETERS
C         X       - ARRAY OF DATA POINTS FOR INDEPENDENT VARIABLE
C         I       - INDEX OF DATA POINTS
C         A       - ARRAY OF PARAMETERS
C         DELTAA - ARRAY OF PARAMETER INCREMENTS
C         NTERMS - NUMBER OF PARAMETERS
C         DERIV  - DERIVATIVES OF FUNCTION
C
C       SUBROUTINES AND FUNCTION SUBPROGRAMS REQUIRED
C         NONE
C
C       MODIFICATIONS FOR FORTRAN II
C         ADD F SUFFIX TO EXP IN STATEMENT 21
C
        SUBROUTINE FDERIV (X, I, A, DELTAA, NTERMS, DERIV)
        DIMENSION X(1), A(1), DELTAA(1), DERIV(1)
     11 XI = X(I)
        Z = (XI - A(2)) / A(3)
        Z2 = Z**2
        IF (Z2 - 50.) 21, 15, 15
     15 DO 16 J=1, 3
     16 DERIV(J) = 0.
        GO TO 24
C
C         ANALYTICAL EXPRESSIONS FOR DERIVATIVES
C
     21 DERIV(1) = EXP (-Z2/2.)
        DERIV(2) = A(1) * DERIV(1) * Z/A(3)
        DERIV(3) = DERIV(2) * Z
     24 DERIV(4) = 1.
        DERIV(5) = XI
        DERIV(6) = XI**2
        RETURN
        END
```

two values of the parameter a_J on either side of the starting point, and the derivative is estimated from the resulting variation in the value of the function.

$$\frac{\partial y(x_I)}{\partial a_J} \simeq \frac{y(a_J + \Delta a_J) - y(a_J - \Delta a_J)}{2\Delta a_J}$$

The input variables are the same as for Program 11-6 except that the parameter increments for computing the derivatives are assumed to be stored in the array DELTAA(J) = Δa_J.

Program 11-7   FDERIV   (Nonanalytical) Derivative of fitting function.

```
C     SUBROUTINE FDERIV (NON ANALYTICAL)
C
C     PURPOSE
C       EVALUATE DERIVATIVES OF FUNCTION FOR LEAST-SQUARES SEARCH
C         FOR ARBITRARY FUNCTION GIVEN BY FUNCTN
C
C     USAGE
C       CALL FDERIV (X, I, A, DELTAA, NTERMS, DERIV)
C
C     DESCRIPTION OF PARAMETERS
C       X      - ARRAY OF DATA POINTS FOR INDEPENDENT VARIABLE
C       I      - INDEX OF DATA POINTS
C       A      - ARRAY OF PARAMETERS
C       DELTAA - ARRAY OF PARAMETER INCREMENTS
C       NTERMS - NUMBER OF PARAMETERS
C       DERIV  - DERIVATIVES OF FUNCTION
C
C     SUBROUTINES AND FUNCTION SUBPROGRAMS REQUIRED
C       FUNCTN (X, I, A)
C         EVALUATES THE FITTING FUNCTION FOR THE ITH TERM
C
C     MODIFICATIONS FOR FORTRAN II
C       NONE
C
      SUBROUTINE FDERIV (X, I, A, DELTAA, NTERMS, DERIV)
      DIMENSION X(1), A(1), DELTAA(1), DERIV(1)
   11 DO 18 J=1, NTERMS
      AJ = A(J)
      DELTA = DELTAA(J)
      A(J) = AJ + DELTA
      YFIT = FUNCTN (X, I, A)
      A(J) = AJ - DELTA
      DERIV(J) = (YFIT - FUNCTN(X,I,A)) / (2.*DELTA)
   18 A(J) = AJ
      RETURN
      END
```

11-5   ERROR DETERMINATION

Since the solution to our least-squares fitting procedure is the result of a search along the χ^2 hypersurface rather than an exact analytical solution, there is no analytical form for the uncertainties σ_{a_j} in the final values of the parameters. We can, however, develop an algorithm which is reasonable and which reduces to the correct analytical expression for linear functions.

The uncertainty in each parameter for a linear least-squares fit was found in Section 8-3 to be related to the diagonal terms in the error matrix ϵ as expressed in Equation (8-28).

$$\sigma_{a_j}{}^2 = \epsilon_{jj} \tag{11-30}$$

In the limit of the approximations we have made, i.e., a parabolic expansion of the χ^2 hypersurface, this uncertainty corresponds to an increase in χ^2 of 1. That is, if we change one parameter a_m by an amount $\Delta a_m = \epsilon_{mm}$, where ϵ is the inverse of the curvature matrix α, and optimize all the other parameters $a_{j \neq m}$ for minimum χ^2, then the new value of χ^2 will be 1 greater than the old value.

$$\chi^2(a_m + \epsilon_{mm}) = \chi^2(a_m) + 1 \tag{11-31}$$

Since the predicted value of χ^2 is the number of data points N minus 1 for each parameter determined from the data,

$$\frac{\chi^2}{N - n} \simeq 1$$

the formula of Equation (11-30) can be considered a reasonable definition of the uncertainty σ_{a_j} for non-linear solutions as well as for linear fitting. Furthermore, this is just the uncertainty we would choose if we trusted the approximations inherent in the expansion of the fitting function.

Variation of χ^2 The proof[1] of Equation (11-31) requires a little sophistication in matrix manipulation. Let us consider the curvature matrix α to be divided into four parts: one is the diagonal element α_{mm} corresponding to the parameter a_m under consideration; another contains all the elements α_{jk} which do not refer to the parameter a_m; and the other two are mirror-image portions of the matrix.

$$\alpha = \begin{bmatrix} \alpha_{jk} & \alpha_{jm} \\ \alpha_{mk} & \alpha_{mm} \end{bmatrix} = \begin{bmatrix} \mathbf{A} & \mathbf{B} \\ \mathbf{B} & \alpha_{mm} \end{bmatrix} \qquad j, k \neq m$$

Let us expand χ^2 in a second-order Taylor's expansion around the minimum where first derivatives are 0.

$$\begin{aligned} \Delta\chi^2 &= \sum_{j=1}^{n} \sum_{k=1}^{n} (\alpha_{jk} \, \Delta a_j \, \Delta a_k) \\ &= (\Delta a \, \mathbf{A} \, \Delta a) + 2(\Delta a_m \, \mathbf{B} \, \Delta a) + (\Delta a_m \, \alpha_{mm} \, \Delta a_m) \end{aligned} \tag{11-32}$$

[1] See Arndt and MacGregor, appendix II.

The matrix notation of Equation (11-32) implies that Δa matrices on the left are row matrices and those on the right are column matrices. The term $\Delta a \, \mathbf{A} \, \Delta a$, for example, comes from multiplying a row matrix Δa with a symmetric matrix $\mathbf{A}$ and multiplying the resulting row matrix with a column matrix Δa. The symmetry of the notation results from the fact that the matrices are symmetric about the diagonal $\alpha_{jk} = \alpha_{kj}$. The matrices Δa are assumed to contain all the elements $\Delta a_{j \neq m}$ except Δa_m.

Let us consider Δa_m to be an increment in the parameter a_m and optimize the fit by choosing appropriate increments for all the other parameters $\Delta a_{j \neq m}$. To do this we set the derivatives of χ^2 with respect to each of these parameter increments equal to 0.

$$\frac{1}{2} \frac{\partial \chi^2}{\partial \Delta a_k} = \sum_{j=1}^{n} (\alpha_{jk} \, \Delta a_j) = (\mathbf{A} \, \Delta a) + (\mathbf{B} \, \Delta a_m) = 0$$
$$\Delta a = -(\mathbf{A}^{-1} \mathbf{B} \, \Delta a_m) \tag{11-33}$$

Substituting Equation (11-33) into Equation (11-32) yields an expression for χ^2 in terms of Δa_m.

$$\Delta \chi^2 = \Delta a_m^2 [\alpha_{mm} - (\mathbf{B}\mathbf{A}^{-1}\mathbf{B})] \tag{11-34}$$

If we consider the error matrix to be divided in the same way,

$$\boldsymbol{\epsilon} = \boldsymbol{\alpha}^{-1} = \left[\begin{array}{c|c} \mathbf{E} & \mathbf{F} \\ \hline \mathbf{F} & \epsilon_{mm} \end{array} \right]$$

then the fact that the error matrix is the inverse of the curvature matrix gives us a relationship between the element ϵ_{mm} and the elements of the curvature matrix.

$$\boldsymbol{\epsilon}\boldsymbol{\alpha} = 1 \Rightarrow \begin{cases} (\mathbf{F}\mathbf{A}) + (\epsilon_{mm}\mathbf{B}) = 0 \\ (\mathbf{F}\mathbf{B}) + (\epsilon_{mm}\alpha_{mm}) = 1 \end{cases}$$
$$\epsilon_{mm}[\alpha_{mm} - (\mathbf{B}\mathbf{A}^{-1}\mathbf{B})] = 1 \tag{11-35}$$

Combining Equations (11-34) and (11-35) yields the result we want.

$$\Delta \chi^2 = \frac{\Delta a_m^2}{\epsilon_{mm}}$$

Independent parameters If the variation of χ^2 with respect to each parameter is independent of the values of the other parameters (at least near the minimum), the curvature and error matrices are diagonal and Equation (11-30) reduces to

$$\sigma_{a_j}{}^2 = \frac{1}{\alpha_{jj}} = \frac{2}{\partial^2 \chi^2 / \partial a_j{}^2} \tag{11-36}$$

In the grid search, the parameters are considered to be independent and the curvature of χ^2 can be determined from the last three points along the search. This method is illustrated in statement 83 of the subroutine GRIDLS of Program 11-1.

In the gradient search, the curvature of χ^2 is never computed with respect to any individual parameter, so evaluation of the uncertainties must be done in addition to the searching.

For analytical solutions, such as those of Sections 11-3 and 11-4, the curvature and error matrices are computed for the fit, and evaluation of the uncertainties follows Equation (11-30). This computation is illustrated in statement 104 of the subroutine CHIFIT and statement 103 of the subroutine CURFIT.

SUMMARY

Non-linear function: One which cannot be expressed as a sum of terms with the parameters appearing only as coefficients of the terms.

Grid search: Vary each parameter in turn, minimizing χ^2 with respect to each parameter independently. Many successive iterations are required to locate the minimum of χ^2 unless the parameters are independent; i.e., the variation of χ^2 with respect to one parameter is independent of the values of the other parameters.

Parabolic interpolation:

$$\Delta a(\min) = a(3) - \Delta a \left[\frac{\chi^2(3) - \chi^2(2)}{\chi^2(3) - 2\chi^2(2) + \chi^2(1)} + \frac{1}{2} \right]$$

Gradient search: Vary all parameters simultaneously, adjusting relative magnitudes of the variations so that the direction of propagation in parameter space is along the direction of steepest descent of χ^2.

Direction of steepest descent: Opposite the gradient $\nabla\chi^2$:

$$(\nabla\chi^2)_i \equiv \frac{\partial\chi^2}{\partial a_j} \simeq \frac{\chi^2(a_j + \Delta a_j) - \chi^2(a_j)}{\Delta a_j}$$

$$\delta a_j = \frac{-\left(\dfrac{\partial\chi^2}{\partial a_j}\Delta a_j{}^2\right)}{\sqrt{\displaystyle\sum_{k=1}^{n}\left(\dfrac{\partial\chi^2}{\partial a_j}\Delta a_j\right)^2}}$$

Parabolic extrapolation of χ^2:

$$\delta a_j = \sum_{k=1}^{n}(\beta_k\epsilon_{jk})$$

$$\beta_k \equiv -\frac{1}{2}\frac{\partial\chi^2}{\partial a_k} \qquad \epsilon \equiv \alpha^{-1} \qquad \alpha \equiv \frac{1}{2}\frac{\partial^2\chi^2}{\partial a_j\,\partial a_k}$$

Linearization of fitting function:

$$\beta_k \equiv \sum\left\{\frac{1}{\sigma_i{}^2}[y_i - y(x_i)]\frac{\partial y(x_i)}{\partial a_k}\right\}$$

$$\alpha_{jk} = \sum\left(\frac{1}{\sigma_i{}^2}\left\{\frac{\partial y(x_i)}{\partial a_j}\frac{\partial y(x_i)}{\partial a_k} - [y_i - y(x_i)]\frac{\partial^2 y(x_i)}{\partial a_j\,\partial a_k}\right\}\right)$$

Gradient-expansion algorithm: Make λ just large enough to insure that χ^2 decreases.

$$\delta a_j = \sum_{k=1}^{n}(\beta_k\epsilon'_{jk})$$

$$\epsilon' = \alpha'^{-1} \qquad \alpha'_{jk} = \begin{cases} \alpha_{jk}(1+\lambda) & \text{for } j = k \\ \alpha_{jk} & \text{for } j \neq k \end{cases}$$

$$\alpha_{jk} \simeq \sum\left[\frac{1}{\sigma_i{}^2}\frac{\partial y(x_i)}{\partial a_j}\frac{\partial y(x_i)}{\partial a_k}\right]$$

Uncertainty in parameter σ_{aj}:

$$\sigma_{aj}{}^2 = \epsilon_{jj} \Rightarrow \Delta\chi^2 = 1$$

EXERCISES

11-1 Derive Equation (11-4).

11-2 Show that the values of the gradients γ_j are dependent on the sizes of the grid metric Δa_j.

CHAPTER TWELVE

FITTING
COMPOSITE CURVES

12-1   AREA DETERMINATION

One task for which non-linear least-squares fitting procedures are well suited is that of determining the position, width, and area under a peak in a spectrum, such as that of Figure 11-5. In many cases the item of greatest interest is the *area* under the peak, which may be a measure of the intensity of a transition or the strength of a reaction. When peaks are not well separated, or where the contribution from background is substantial, least-squares fitting can provide a consistent method of extracting such information from data.

The method of least squares is considered to be an unbiased

estimator of the fitting parameters; i.e., since the procedure minimizes the sample variance s^2 of the data from the fitted curve with respect to each of the parameters, all parameters are presumed to be estimated as well as possible, considered independently. The assumption of unbiased estimate is based on the validity of the least-squares method and the fitting function in describing the data. If we try to fit data with an incorrect fitting function, or try to fit to data which do not follow the Gaussian distribution, such a fitting procedure may not yield the optimum results.

Area under a peak for Poisson distribution   Curiously enough, if the data are distributed around each data point according to a Poisson distribution, as for a counting experiment, the method of least squares consistently *underestimates* the area under a peak by an amount approximately equal to the value of χ^2. To show this, let us consider fitting such data with a Gaussian peak plus a polynomial background as in Example 11-1

$$y(x) = ae^{-bx^2} + c \tag{12-1}$$

where we have simplified the function to three parameters for clarity.

Using the method of least squares, we define χ^2 to be the weighted sum of squares of deviations of the data from the fitted curve

$$\chi^2 = \sum \left[\frac{1}{\sigma_i{}^2} (y_i - ae^{-bx^2} - c)^2 \right] \tag{12-2}$$

and minimize χ^2 with respect to each of the parameters a, b, and c simultaneously.

$$-\frac{1}{2} \frac{\partial \chi^2}{\partial a} = \sum \left[\frac{1}{\sigma_i{}^2} (y_i - ae^{-bx^2} - c)e^{-bx^2} \right] = 0$$

$$-\frac{1}{2} \frac{\partial \chi^2}{\partial c} = \sum \left[\frac{1}{\sigma_i{}^2} (y_i - ae^{-bx^2} - c) \right] = 0 \tag{12-3}$$

The value of χ^2 at the minimum is obtained by substituting Equations (12-3) into Equation (12-2).

$$\chi^2_{\min} = \sum \left[\frac{1}{\sigma_i^2} \left(y_i - ae^{-bx^2} - c \right) y_i \right] + \frac{1}{2} \left(a \frac{\partial \chi^2}{\partial a} + c \frac{\partial \chi^2}{\partial c} \right)$$

$$= \sum \left[\frac{y_i}{\sigma_i^2} \left(y_i - ae^{-bx^2} - c \right) \right] \tag{12-4}$$

If the data represent the number of counts per unit time in a detector, then they are distributed according to the Poisson distribution and we can approximate $\sigma_i^2 \simeq y_i$. Equation (12-4) becomes

$$\chi^2_{\min} \simeq \Sigma y_i - \Sigma y(x_i) = \text{area (data)} - \text{area (fit)}$$

which shows that the area under the total fit is underestimated by an amount equal to $\chi^2_{\min}$.

For this derivation we require only that the fitting function consist of a sum of terms, each one of which is multiplied by a coefficient.

$$y(x) = \sum_{j=1}^{n} [a_j f_j(x)]$$

The functions $f_j(x)$ can contain any number of other parameters in non-linear form, but may not contain any of the coefficients a_j. Even reparameterizing the function of Equation (12-1) to minimize χ^2 with respect to the area A explicitly

$$y(x) = A \sqrt{\frac{b}{\pi}} \, e^{-bx^2} + c$$

will not affect the discrepancy between the actual and estimated areas.

Note that if the data were distributed with a constant uncertainty $\sigma_i = \sigma$, the second equation of Equations (12-3) is sufficient to ensure that $\Sigma y(x_i) = \Sigma y_i$. It is the assumption of a Poisson distribution for the data $\sigma_i^2 \simeq y_i$ which yields the discrepancy between the actual and estimated areas. But, if the fit to the data is perfect, $\chi^2 = 0$ and the estimated and actual areas are equal.

Reducing χ^2 It is clear then that for any analytical method of least-squares fitting to data which follow a Poisson distribution, the optimum estimate of the area should be taken as the sum of the least-squares estimate plus the value of χ^2 at the minimum. If the region of fitting includes background as well as a peak, however, the discrepancy is summed over the whole range of the fitting and this correction is too large.

One obvious alternative is to reduce the value of χ^2 at the minimum so that the correction is small. A method of accomplishing this reduction, which is not universally accepted but which is well justified from practical considerations, is the technique of smoothing the data, as described in Section 13-1. Under any smoothing process there can be no overall gain in information, and a net improvement in the fit to the area must be offset by an increased uncertainty in the estimation of other parameters, such as the width of the peak. But smoothing will decrease the value of χ^2 at the minimum, often by nearly an order of magnitude, and thereby reduce the uncertainty in the estimation of the area.

The same concepts can be applied to the optimized estimation of any one of the parameters of a fitting function. What is required is to minimize χ^2 by a technique which does not alter the representation of that parameter in the data. Smoothing data by averaging over adjacent channels, for example, does not change the sum of the number of counts in the range of channels.

Similarly, if the fitting function is not a precise description of the data (e.g., a Gaussian peak used to fit data with a tail or a Lorentzian shape), the estimated area may be incorrect with a discrepancy relative to the value of χ^2. If it is not convenient to improve the analytical form of the fitting function, it may be preferable to manipulate the data by a smoothing procedure to improve the estimate of the area under the data by decreasing the minimum value of χ^2. This is true even if the data have constant uncertainties if the fitting function consists of a peak superimposed on a background. The error results in an incorrect separation of peak and background.

12-2 BACKGROUND SUBTRACTION

If the fitting function $y(x)$ is one which is separable into a peak $y_p(x)$ plus a background $y_b(x)$,

$$y(x) = y_p(x) + y_b(x) \tag{12-5}$$

as in Equation (12-1) or Example 11-1, it is often convenient to consider at least some facets of the fitting procedure separately.

The method of least squares procedure of minimizing χ^2 with respect to each of the parameters a_j

$$\frac{\partial}{\partial a_j} \sum \left\{ \frac{1}{\sigma_i^2} [y_i - y_p(x_i) - y_b(x_i)]^2 \right\} = 0 \tag{12-6}$$

can be considered equally well in terms of fitting the sum of the curves $y(x)$ to the total yield y_i or of fitting one function $y_p(x)$ to the difference spectrum $y_i' = y_i - y_b(x_i)$. The only provision is that the uncertainties in the data points $\sigma_i' = \sigma_i$ must be the same in both calculations.

Fitting region The general procedure for fitting a function such as that of Equation (12-5) is to fit both the peak function $y_p(x)$ and the background function $y_b(x)$ simultaneously over the entire region of interest. However, if the background curve can be assumed to be a slowly varying function under the peak, as in Figure 11-5, and may reasonably be interpolated under the peak from fitting on both sides, it is often preferable to fit the background curve $y_b(x)$ outside the region of the peak and to fit the peak function $y_p(x)$ only in the region of the peak.

The reason for this separation of fitting regions is that χ^2 measures not only the deviation of the parameters a_j from an ideal fit but also the discrepancy between the form chosen for the fitting function $y(x)$ and the parent distribution of the data. If the shape of the fitting function does not simulate that of the parent distribution exactly (for example, if the peak function $y_p(x)$ is taken to be symmetric about its centroid while the peaks in the data spectrum contain tails on one side), the value of χ^2 may be influenced considerably by the region where the peak curve merges with the background.

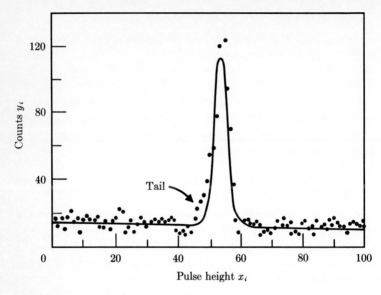

FIGURE 12-1 Illustration of danger of fitting data with a function which incorrectly describes parent population. Note that fluctuations of data are small compared with discrepancies in the region of the tail of the peak on the left side. Largest contribution of χ^2 comes from this region.

Such a mismatch is indicated in Figure 12-1 where the peak of the data exhibits a tail on the low side. For a reasonable fit to the entire region, the largest contribution to the value of χ^2 comes from this tail where the predicted uncertainties of the data are small and the actual deviations are large.

This situation is aggravated when the number of counts in the peak is large, if we assume Poisson statistics. For this case, the expected deviations σ_i are small fractions of the observed counts y_i and discrepancies between the theoretical and parent distributions will be most apparent. If the number of counts is small and the σ_i are sizable fractions of the y_i, the deviations due to discrepancies in the distributions will be masked by statistical fluctuations.

In fitting the peak and background functions over different parts of the spectrum, however, it is important to note that the fit is still to the function $y(x)$ of Equation (12-5); that is, in the region outside the peak where the background is being fitted, the tail of the peak must be included and vice-versa. Similarly, the value of χ^2 to be minimized is still that of Equation (12-6), including the total uncertainties σ_i in each data point. The only difference between this method and fitting both functions over the whole region is that the parameters of only one function are optimized in each region.

Error determination   The uncertainties σ_{a_j} of the values of the parameters determined from least-squares fits were discussed in Section 11-5. These uncertainties resulted from the ambiguities of fitting curves to data points which fluctuated about their expected values. If, either by chance or by probability, the data points fall very close to a smooth curve, the value of χ^2 can be very small. This means that the fit describes the *data* very well, but the description of the *parent distribution* may not be so precise. The uncertainties σ_{a_j} in the parameters reflect the uncertainties in the data points as samples of the parent population as well as the uncertainties of describing the data.

As long as the value of χ^2 is reasonably small ($\chi_\nu^2 \lesssim 1$), the uncertainty in each parameter is essentially that of the parent distribution. For example, if the least-squares fitting is rigorously valid, the estimated area under a curve is equal to the sum of the data points regardless of the value of χ^2. Thus, the uncertainty in the area is the uncertainty of the distribution of this sum, which can be considered independently of the precision of the fit. An extremely small value of χ^2 is assumed to result from purely statistical fluctuations if the values of the σ_i can be trusted.

If χ^2 is unreasonably large, the presumption is usually that the data are scattered more violently than statistics would predict, and, therefore, the uncertainty in the parameters is larger than the corresponding uncertainties of the parent distribution. The formulas for estimating the uncertainties of parameters in the preceding chapters take all these considerations into account.

Area determination In the case of area determinations, the uncertainty σ_A can often be estimated most easily from consideration of the uncertainty in the parent distribution. If the data are distributed according to the Poisson distribution, the uncertainty in the area A may be estimated according to the approximation of Equation (6-16).

$$\sigma_A^2 \simeq A$$

If the parameter being estimated is the area under a peak which is superimposed on a background (as in Example 11-1), the area to which the fit is made is effectively a difference between the total area under the total curve $y(x)$ and the area under the background curve $y_b(x)$. Thus, the statistical uncertainty in the peak area σ_{A_p} is the statistical error of a difference (see Section 4-2).

$$\sigma_{A_p}^2 = \sigma_A^2 + \sigma_{A_b}^2 \simeq A + A_b$$

If $\chi_\nu^2 > 1$, as a rough rule multiply the uncertainty by the reduced chi-square.

$$s_A^2 \simeq \sigma_A^2 \chi_\nu^2$$

SUMMARY

Area under a curve with Poisson statistics:

$$\chi_{\min}^2 \simeq \sum \left\{ \frac{y_i}{\sigma_i^2} [y_i - y(x_i)] \right\} \simeq \text{area (data)} - \text{area (fit)}$$

Background subtraction:

$$y_p(x) = y(x) - y_b(x)$$
$$y_i' \equiv y_i - y_b(x_i)$$
$$\frac{\partial}{\partial a_j} \sum \left\{ \frac{1}{\sigma_i^2} [y_i' - y_b(x_i)]^2 \right\} = 0$$

where σ_i is the total uncertainty in the data point ($\sigma_i^2 \simeq y_i$).

Area determination:

$$\sigma_{A_p}^2 = \sigma_A^2 + \sigma_{A_b}^2 \simeq A + A_b$$

Uncertainty in area for $\chi_\nu^2 > 1$:

$$s_A^2 \simeq \sigma_A^2 \chi_\nu^2$$

CHAPTER THIRTEEN

DATA MANIPULATION

The concept of smoothing data is not one which meets with universal approval. The discussion which follows should be considered with the admonition of a *caveat emptor:* for rigorously valid least-squares fitting, smoothing is neither desirable nor permissible; however, there are cases where smoothing can be beneficial, and, therefore, the techniques are introduced.

Consider, for example, the discussion of Section 12-1. Least-squares fitting techniques applied to data which are distributed according to Poisson distributions, rather than Gaussian distributions, yield incorrect results which can be improved by decreasing the minimum value of χ^2. Similarly, if the shape of the fitting function does not exactly simulate that of the parent distribution, a better fit to the data by decreasing χ^2 can yield an improved estimate of the area under a peak.

In other words, if the least-squares fitting procedure is not rigorously valid but represents a convenient approximation to the proper procedure, it may be possible to estimate some of the parameters of a fitting function better by applying some technique, such as smoothing the data, which yields a smaller minimum value of χ^2. The improvement in the estimate of one parameter must, of course, be accompanied by a decrease in information of some other parameter or parameters. For example, an improved estimate of the area under a peak as in Section 12-1 was accompanied by an increased uncertainty in the estimates of the width and position of the peak.

It is not possible to describe the parent distribution better than the uncertainty associated with the data. The data represent a random sample from the parent population, and an exact description of the data merely represents an estimate of the parent distribution. When applied rigorously, the least-squares method provides the most unbiased estimate of each of the parameters with errors which approximate the uncertainties from the parent distribution. But when the least-squares method is not rigorously valid, the resulting parameters may be inaccurate descriptions of the data. In such a case, it is the errors associated with these inaccuracies which may be minimized by manipulating the data.

Of course, whatever smoothing or other manipulation is done must conserve the information pertaining to the desired parameter. The averaging techniques discussed below, for example, conserve the area under a peak but not the width of the peak. Similarly, this method would be useful for improving the estimate of the constant term of a polynomial but not the coefficients of the other terms.

Area under a Gaussian peak   Let us consider the problem of Section 12-1 where we wanted to find the area under a Gaussian peak superimposed on a slowly varying background. Let us assume that the background slope is gentle enough that smoothing will not affect its determination drastically.

We can approximate the Gaussian peak with a binomial distribution with $p = \frac{1}{2}$ (see Section 3-1).

$$y(x) = \frac{1}{\sigma \sqrt{2\pi}} e^{-\frac{1}{2}\left(\frac{x-\mu}{\sigma}\right)^2} \simeq \left(\frac{1}{2}\right)^n \frac{n!}{X!(n-X)!} \tag{13-1}$$

We can relate the widths σ and the means of the two distributions

$$\sigma_B{}^2 = \frac{n}{4} = \sigma^2 \qquad \bar{X} = \frac{1}{2}n \qquad \bar{x} = \mu$$

to find the relationships between the parameters.

$$n = 4\sigma^2 \qquad X = x - \mu + 2\sigma^2 \tag{13-2}$$

Let us smooth the data by averaging over adjacent channels with a binomial distribution spanning three channels.

$$y'(x) = \frac{1}{4}y(x-1) + \frac{1}{2}y(x) + \frac{1}{4}y(x+1) \tag{13-3}$$

If we fold this averaging into the distribution of Equation (13-1), the result is also binomial.

$$y(x) = \left(\frac{1}{2}\right)^n \frac{n!}{(\frac{1}{2}n + x - \mu)!(\frac{1}{2}n - x + \mu)!}$$

$$y'(x) = \left(\frac{1}{2}\right)^{n+2} \frac{(n+2)!}{(\frac{1}{2}n + 1 + x - \mu)!(\frac{1}{2}n + 1 - x + \mu)!}$$

The new distribution has the same mean $\bar{x} = \mu$ but a larger width $\sigma'^2 = \frac{1}{4}n' = \frac{1}{4}(n+2)$, with the variance increased by $\frac{1}{2}$.

$$\sigma'^2 = \sigma^2 + \frac{1}{2} \tag{13-4}$$

Similarly, we could smooth over five channels by using a formula similar to Equation (13-3) but with five terms whose coefficients are those of the binomial expansion.

$$y''(x) = \frac{1}{16}y(x-2) + \frac{1}{4}y(x-1) + \frac{3}{8}y(x) \\ + \frac{1}{4}y(x+1) + \frac{1}{16}y(x+2)$$

A five-channel smoothing is identical to two successive smoothings over three channels and yields a variance which is increased accordingly $\sigma''^2 = \sigma^2 + 1$. Any such smoothing over $2n + 1$ adjacent channels is equivalent to n smoothings over three channels.

If we apply the smoothing of Equation (13-3) to a Gaussian distribution, the resulting distribution will also be nearly Gaussian because the shapes of the binomial and Gaussian distributions are nearly alike. In fact, if we are applying the smoothing because the original shape is not Gaussian enough, the averaging may make the shape even more nearly Gaussian.

If the width of the original Gaussian is not too small ($\sigma > 1$), the increase of Equation (13-4) should not be drastic because of the addition by quadrature. For a width of $\sigma = 2$, for example, the new width $\sigma' = 2.1$ is only 5% larger.

If the original width is very small ($\sigma < 1$), the approximation of Equation (13-1) is not valid because the Gaussian and binomial distributions are only similar in the limit of large n. A Gaussian fit to the data without smoothing would not be valid either, however, because the parameters of the fit are only meaningful if $\sigma \geq 1$. Since the averaging itself is a binomial distribution, the result is still expected to be a better approximation to a Gaussian distribution than the original data. For a smoothing over three channels, a Gaussian fit requires $\sigma \geq \sqrt{\frac{1}{2}}$ for the original data.

EXAMPLE 13-1 The data of Example 11-1 are graphed in Figure 13-1 after one three-channel smoothing. Note that the fluctuations of the data around the fitted curve are considerably smaller, but there are gross structures averaging over several channels which are more apparent. The structures are not introduced by the smoothing, but they are more noticeable because the channel-to-channel fluctuations which masked them are removed.

The estimated areas under the peak and background after smoothing are smaller than before smoothing, with a difference on the order of the χ^2. The estimated values after smoothing are more accurate because the least-squares estimate of the area is underestimated by χ^2.

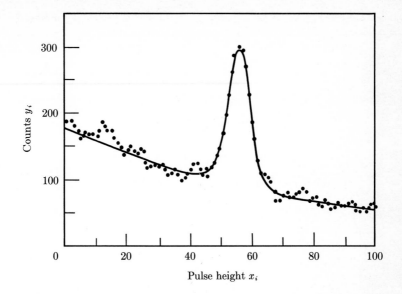

FIGURE 13-1 Data and fit of Example 11-1 after smoothing. Compare the fluctuations of these data about the fit with those of Figure 11-5.

Program 13-1 The smoothing procedure of Equation (13-3) is illustrated in the computer routine SMOOTH of Program 13-1. This is a Fortran subroutine to smooth an array of data with an average over groups of three adjacent channels. The input variables are Y and NPTS. The data to be averaged are assumed to be stored in the array $Y(I) = y_i$ with NPTS $= N$ as the number of data points.

The averaging is iterated in the DO loop of statements 21–24, using YI as the contents of $Y(I - 1)$ before averaging. The two end channels are smoothed over only two channels, as if the next channel had the same contents.

$$y_1 = \tfrac{3}{4}y_1 + \tfrac{1}{4}y_2$$

13-2 INTERPOLATION AND EXTRAPOLATION

One reason for fitting the data as we have is to extract from the data values for parameters which describe the parent popu-

Program 13-1 **SMOOTH** Smoothing procedure.

```
C       SUBROUTINE SMOOTH
C
C       PURPOSE
C         SMOOTH A SET OF DATA POINTS BY AVERAGING ADJACENT CHANNELS
C
C       USAGE
C         CALL SMOOTH (Y, NPTS)
C
C       DESCRIPTION OF PARAMETERS
C         Y       - ARRAY OF DATA POINTS
C         NPTS    - NUMBER OF DATA POINTS
C
C       SUBROUTINES AND FUNCTION SUBPROGRAMS REQUIRED
C          NONE
C
C       MODIFICATIONS FOR FORTRAN II
C          NONE
C
        SUBROUTINE SMOOTH (Y, NPTS)
        DIMENSION Y(1)
11      IMAX = NPTS - 1
        YI = Y(1)
21      DO 24 I=1, IMAX
        YNEW = (YI + 2.*Y(I) + Y(I+1)) / 4.
        YI = Y(I)
24      Y(I) = YNEW
25      Y(NPTS) = (YI + 3.*Y(NPTS)) / 4.
        RETURN
        END
```

lation from which the data were derived. Another reason is to evaluate the function which the data represent. By fitting the data with an analytical function, we can determine the value at intermediate points or use information from several data points to determine the value at one point more precisely.

Under some circumstances, when the value of a function $y(x)$ is desired in a region for which measurements y_i at points x_i have been made, it is preferable to evaluate $y(x)$ approximately by interpolation or extrapolation from the values of y_i directly, rather than by determining an analytical function for $y(x)$.

Similarly, some functions which have complex analytical expressions can be evaluated more simply for computation by interpolation between tabulated values. For example, probability functions for χ^2 or F tests, as discussed in Chapter 10, are more easily treated in this manner.

We will develop a technique for interpolating between values of data points y_i which correspond to equally spaced values of the independent variable x_i. Later we will generalize the method to include data which are not equally spaced.

Uniform spacing Consider N pairs of data points (x_i, y_i) evenly spaced (in x_i) and ordered such that x_i varies linearly with i. We are given a value of x and asked to find the corresponding value of y by interpolation. Let us designate one pair of these data points as (x_1, y_1). This pair need not be the first pair, but instead should be one with a value of x_1 near the given value of x. Our method of interpolation will be to fit n pairs of data points ($i = 1$ to n with the above definition for $i = 1$) with a power-series polynomial of degree $n - 1$ and solve this polynomial $y(x)$ to find the value of y corresponding to the given value of x.

Consider a first-degree polynomial fit

$$y_1(x) = a_1 + b_1 \Delta \tag{13-5}$$

where Δ is a dimensionless quantity which expresses the difference between x and x_1 in terms of the spacing between the values of the x_i.

$$\Delta \equiv \frac{x - x_1}{x_2 - x_1} \tag{13-6}$$

We can determine a_1 and b_1 by equating the values of $y_1(x)$ at x_1 and x_2 to y_1 and y_2.

$$a_1 = y_1 \qquad b_1 = y_2 - y_1$$

Similarly, if we consider a quadratic polynomial fit to three data points,

$$y_2(x) = a_2 + b_2 \Delta + c_2 \Delta^2$$
$$a_2 = a_1 \qquad b_2 = b_1 - c_2 \qquad c_2 = \tfrac{1}{2}(y_3 - 2y_2 + y_1)$$

we can express the polynomial in terms of the parameters of Equation (13-5).

$$y_2(x) = a_1 + b_1 \Delta + c_2 \Delta (\Delta - 1) \tag{13-7}$$

By considering a third-degree polynomial, we can determine a general expression for an $(n - 1)$th-degree polynomial with n terms.

$$y_3(x) = a_3 + b_3\Delta + c_3\Delta^2 + d_3\Delta^3$$
$$= a_1 + b_1\Delta + c_2\Delta(\Delta - 1) + d_3\Delta(\Delta - 1)(\Delta - 2)$$

$$a_3 = a_1 \qquad\qquad b_3 = b_1 - c_3 - d_3$$
$$c_3 = c_2 - 3d_3 \qquad d_3 = \tfrac{1}{6}(y_4 - 3y_3 + 3y_2 - y_1)$$

Therefore, we can fit n adjacent data points exactly with an $(n - 1)$th-degree polynomial of the form

$$y(x) = a_1 + a_2\Delta + a_3\Delta(\Delta - 1) + a_4\Delta(\Delta - 1)(\Delta - 2)$$
$$+ \cdots$$
$$= a_1 + \sum_{j=2}^{n}\left[a_j \prod_{i=1}^{j-1}(\Delta - i + 1)\right] \qquad (13\text{-}8)$$

where the coefficients a_j have forms which are similar to terms in a binomial expansion.

$$a_j = \sum_{i=1}^{j}\left[(-1)^{j-i}\frac{y_i}{(i - 1)!(j - i)!}\right] \qquad (13\text{-}9)$$

$$a_1 = y_1 \qquad\qquad a_2 = y_2 - y_1$$
$$a_3 = \tfrac{1}{2}(y_3 - 2y_2 + y_1) \qquad a_4 = \tfrac{1}{6}(y_4 - 3y_3 + 3y_2 - y_1)$$
$$a_5 = \tfrac{1}{24}(y_5 - 4y_4 + 6y_3 - 4y_2 + y_1)$$

EXAMPLE 13-2 The meaning of the terms in the expansion of Equation (13-8) is illustrated in Figure 13-2. Consider three data points y_1, y_2, and y_3 corresponding to x_1, x_2, and x_3. We wish to evaluate y at x with the quadratic fit of Equation (13-7).

The first term is just the value of y_1. The second term adds a correction for a linear interpolation, and the third term corrects for the curvature in the parabolic dashed curve. Note that if $\Delta = 0$, the formula reduces to $y(x_1) = y_1$. Similarly, if $\Delta = j$, the formula yields $y(x_j) = y_j$, but there are j terms in the sum. For $j = 3$, for example,

$$y(x_3) = a_1 + 2(a_2 + a_3) = y_3$$

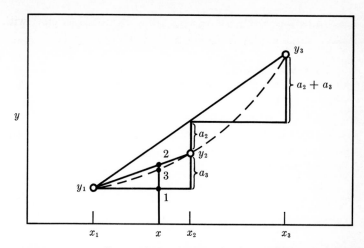

FIGURE 13-2 Comparison of terms for interpolation between data points with quadratic polynomial. Points marked 1, 2, and 3 signify the value of the interpolation after adding, respectively, the first, second, and third terms of the expansion. Dashed curve is the final parabolic fit.

and the geometrical construction in Figure 13-2 shows how this equality comes about.

Extrapolation The formula of Equations (13-8) and (13-9) is perfectly general for fitting n successive equally spaced data points exactly with an $(n - 1)$th-degree polynomial. The position of the first data point (x_1, y_1) can be anywhere so far as the derivation is concerned. For the optimum interpolation, however, the values of x_1 and x_n should straddle the value of x and be nearly equally distant from it.

The same formula can be used for extrapolation of values beyond the region of the data, but the uncertainty in the validity of the approximation increases as x gets farther from the average of x_1 and x_n. The approximation includes both the degree of the polynomial used in the extrapolation and the uncertainties in the coefficients of the polynomial resulting from fluctuations in the data. As a rough rule for extrapolation, the degree n should be at least as large as the deviation Δ from the nearest data point.

Variable spacing We can generalize the results of the preceding discussion to include the case of unequally spaced values of x_i. As before, we define a dimensionless deviation of x from x_1 (with reference to $x_2 - x_1$) similar to that of Equation (13-6), and we define the spacings of other points x_i in the same manner.

$$\Delta \equiv \frac{x - x_1}{x_2 - x_1} \qquad \Delta_i \equiv \frac{x_i - x_1}{x_2 - x_1} \tag{13-10}$$

The general formula for an $(n - 1)$th-degree polynomial fit to n successive data points is very similar to that of Equation (13-8).

$$
\begin{aligned}
y(x) &= a_1 + a_2\Delta + a_3\Delta(\Delta - \Delta_2) \\
&\qquad\qquad + a_4\Delta(\Delta - \Delta_2)(\Delta - \Delta_3) + \cdots \\
&= a_1 + \sum_{j=2}^{n} \left[a_j \prod_{i=1}^{j-1} (\Delta - \Delta_i) \right]
\end{aligned} \tag{13-11}
$$

The definitions of the coefficients are considerably more complicated.

$$a_k = \frac{y_k}{\displaystyle\prod_{i=1}^{k-1} (\Delta_k - \Delta_i)} - \sum_{j=1}^{k-1} \frac{a_j}{\displaystyle\prod_{i=j}^{k-1} (\Delta_k - \Delta_i)} \tag{13-12}$$

$$a_1 = y_1 \qquad a_2 = \frac{1}{\Delta_2} (y_2 - a_1)$$

$$a_3 = \frac{y_3 - a_2\Delta_3 - a_1}{\Delta_3(\Delta_3 - \Delta_2)}$$

$$a_4 = \frac{y_4 - a_3\Delta_4(\Delta_4 - \Delta_2) - a_2\Delta_4 - a_1}{\Delta_4(\Delta_4 - \Delta_2)(\Delta_4 - \Delta_3)}$$

The formula of Equations (13-10) to (13-12) reduces to that of Equations (13-8) and (13-9) for the special case of $\Delta_i = i - 1$ which corresponds to equally spaced values of x_i.

Program 13-2 The calculations of the formula of Equations (13-11) and (13-12) are illustrated in the computer routine

INTERP of Program 13-2. This is a Fortran subroutine to make an $(n - 1)$th-degree polynomial fit to pairs of data points (x_i, y_i) ordered for monotonically increasing but not necessarily equally spaced x_i, and to interpolate (or extrapolate) to find the value of y corresponding to a given value of x. The input variables are X, Y, NPTS, NTERMS, and XIN, and the output variable is YOUT.

The pairs of data points are assumed to be stored in the arrays $X(I) = x_I$ and $Y(I) = y_I$ with NPTS $= N$ data pairs in all. The variable NTERMS $= n$ is the number of terms in the polynomial which is one more than the degree. The variable XIN $= x$ represents the given value of the independent variable, and the variable YOUT $= y$ represents the result of the interpolation.

The program searches in statements 11–27 to find the appropriate value of x_i to use for x_1. If possible, x_1 is chosen so that x_1 and x_n straddle x. If the value of x is too close to the lower or upper limit of the values of x_i, however, the corresponding value of x_1 or x_n is set equal to the limiting value.

The dimensionless deviations DELTAX $= \Delta$ and DELTA(I) $= \Delta_I$ are evaluated in statements 31–35 according to Equations (13-10). The coefficients $A(J) = a_J$ of Equations (13-12) are accumulated in statements 40–50, and the terms in the expansion of Equation (13-11) are evaluated and summed in statements 51–57. The variable YOUT $= y$ is set equal to the sum in statement 60 and returned to the main program as the result of the interpolation. If, by chance, the given value of x is equal to one of the data values of x_i, the interpolation is bypassed in statements 17–18, setting YOUT $= y = y_i$.

13-3 AREA INTEGRATION

As in Section 13-2, there are circumstances when it is preferable to evaluate the area under a curve corresponding to a set of data points by summing over the data points themselves instead of fitting them and extracting an area parameter from the fit. This is especially true if the background contribution is small or has been compensated for. The total cross section for a reaction,

Program 13-2 INTERP Interpolation and extrapolation of data.

```
C     SUBROUTINE INTERP
C
C     PURPOSE
C       INTERPOLATE BETWEEN DATA POINTS TO EVALUATE A FUNCTION
C
C     USAGE
C       CALL INTERP (X, Y, NPTS, NTERMS, XIN, YOUT)
C
C     DESCRIPTION OF PARAMETERS
C       X       - ARRAY OF DATA POINTS FOR INDEPENDENT VARIABLE
C       Y       - ARRAY OF DATA POINTS FOR DEPENDENT VARIABLE
C       NPTS    - NUMBER OF PAIRS OF DATA POINTS
C       NTERMS  - NUMBER OF TERMS IN FITTING POLYNOMIAL
C       XIN     - INPUT VALUE OF X
C       YOUT    - INTERPOLATED VALUE OF Y
C
C     SUBROUTINES AND FUNCTION SUBPROGRAMS REQUIRED
C       NONE
C
C     MODIFICATIONS FOR FORTRAN II
C       OMIT DOUBLE PRECISION SPECIFICATIONS
C
C     COMMENTS
C       DIMENSION STATEMENT VALID FOR NTERMS UP TO 10
C       VALUE OF NTERMS MAY BE MODIFIED BY THE PROGRAM
C
```

for example, can be extracted from data for an angular distribution by integrating the distribution as a function of angle. The integration can be accomplished analytically or by finding the total area under a measured angular distribution.

The simplest form of area integration is just the sum of all data points (normalized to the same units and equal spacing). But if we believe that there is additional information to be gained from the shape of the functional behavior of the data, we might find a weighted sum which would yield a more accurate estimate of the expected area.

Generally, we expect random fluctuations of the data to average out, at least in the middle of the range of the data points so that the sum of data points has better relative precision than each individual point. Furthermore, we presume that, at least in the middle of the range, no one data point is better than the one next to it, so our technique for evaluating areas must treat all such data points equally. Near the edge of the range of data points, however, these assumptions break down because the number of neighboring points over which to average becomes fewer. In

Program 13-2 INTERP *(continued)*

```
      SUBROUTINE INTERP (X, Y, NPTS, NTERMS, XIN, YOUT)
      DOUBLE PRECISION DELTAX, DELTA, A, PROD, SUM
      DIMENSION X(1), Y(1)
      DIMENSION DELTA(10), A(10)
C
C         SEARCH FOR APPROPRIATE VALUE OF X(1)
C
   11 DO 19 I=1, NPTS
      IF (XIN - X(I)) 13, 17, 19
   13 I1 = I - NTERMS/2
      IF (I1) 15, 15, 21
   15 I1 = 1
      GO TO 21
   17 YOUT = Y(I)
   18 GO TO 61
   19 CONTINUE
      I1 = NPTS - NTERMS + 1
   21 I2 = I1 + NTERMS - 1
      IF (NPTS - I2) 23, 31, 31
   23 I2 = NPTS
      I1 = I2 - NTERMS + 1
   25 IF (I1) 26, 26, 31
   26 I1 = 1
   27 NTERMS = I2 - I1 + 1
C
C         EVALUATE DEVIATIONS DELTA
C
   31 DENOM = X(I1+1) - X(I1)
      DELTAX = (XIN - X(I1)) / DENOM
      DO 35 I=1, NTERMS
      IX = I1 + I - 1
   35 DELTA(I) = (X(IX) - X(I1)) / DENOM
C
C         ACCUMULATE COEFFICIENTS A
C
   40 A(1) = Y(I1)
   41 DO 50 K=2, NTERMS
      PROD = 1.
      SUM = 0.
      IMAX = K - 1
      IXMAX = I1 + IMAX
      DO 49 I=1, IMAX
      J = K - I
      PROD = PROD * (DELTA(K) - DELTA(J))
   49 SUM = SUM - A(J)/PROD
   50 A(K) = SUM + Y(IXMAX)/PROD
C
C         ACCUMULATE SUM OF EXPANSION
C
   51 SUM = A(1)
      DO 57 J=2, NTERMS
      PROD = 1.
      IMAX = J - 1
      DO 56 I=1, IMAX
   56 PROD = PROD * (DELTAX - DELTA(I))
   57 SUM = SUM + A(J)*PROD
   60 YOUT = SUM
   61 RETURN
      END
```

particular, the first and last data points in a range may be considered to be more suspect because the next data points are not included in the integration.

A number of techniques have been proposed to optimize the method of weighting data for area integration. Some of these weight alternate data points throughout the range with different weights. For example, Simpson's rule for parabolic interpolation between alternate data points yields a formula which weights half the points twice as much as the rest.

$$\text{Area (Simpson)} = \tfrac{1}{3}(y_1 + 4y_2 + 2y_3 + 4y_4 + \cdots + y_N)$$

We will forgo any discussion of such techniques on the assumption that we have no criteria for choosing one group of data points over another except for edge effects near the limit of the range.

Linear interpolation   Inherent in the concept of summing data points to effect an area integration is the assumption that the area can be approximated by interpolating linearly between data points. If we have N equally spaced data points, however, the area is equal to the average value between data points times the spacing summed over $N - 1$ intervals. If we apply the linear interpolation rigorously, we see that the two end data points should only be weighted by half as much as the rest.

$$\text{Area (linear)} = \tfrac{1}{2}(y_1 + y_N) + \sum_{i=2}^{N-1} y_i$$

Since a linear interpolation is probably not a very precise description of the behavior of the function, we might consider using a higher-degree polynomial to describe the variation between data points, averaging over information from more than the two data points bracketing each interval. A convenient method of doing so is to fit an $(n-1)$th-degree polynomial exactly to n data points straddling each interval, and to use the analytical form for the area under the polynomial to determine the area in the interval.

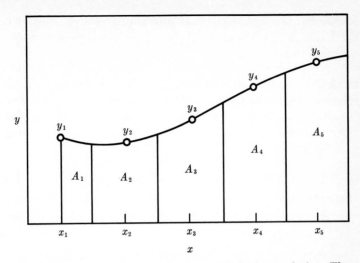

FIGURE 13-3 Area integration by parabolic interpolation. The intervals are chosen symmetrically about the data points. Note that the parabolic interpolation for the first interval is not symmetrical.

Parabolic interpolation Consider the data points shown in Figure 13-3. We wish to fit these data in the region of each interval with a quadratic polynomial and use the analytical expression for that polynomial to evaluate the area A_i in that interval. Since the number of data points used in fitting each parabola is odd, the intervals must be chosen straddling data points, rather than between them, to insure symmetrical fitting.

The coefficients for a quadratic fit to three data points can be evaluated from the formulae of Section 13-2.

$$y(x) = a + bx + cx^2 \tag{13-13}$$
$$a = y_1 \qquad b = 2y_2 - \tfrac{1}{2}(y_3 + 3y_1) \qquad c = \tfrac{1}{2}(y_3 - 2y_2 + y_1)$$

The area in the first interval A_1, which is only half as wide as the others, is determined by integrating Equation (13-13).

$$A_1 = [ax + \tfrac{1}{2}bx^2 + \tfrac{1}{3}cx^3]_0^{\frac{1}{2}} = \tfrac{1}{24}(8y_1 + 5y_2 - y_3)$$

Similarly, the area in the next interval is

$$A_2 = [ax + \tfrac{1}{2}bx^2 + \tfrac{1}{3}cx^3]_{\frac{1}{2}}^{\frac{3}{2}}$$
$$= \tfrac{1}{24}(y_1 + 22y_2 + y_3) \tag{13-14}$$

which is heavily weighted in favor of the middle data point. The areas of the following intervals A_i are evaluated in the same way as that of A_2, with the indices of Equation (13-14) incremented for each successive interval.

$$A_i = \tfrac{1}{24}(y_{i-1} + 22y_i + y_{i+1}) \qquad i = 2, N - 2$$

If we add these areas together, we find that in the middle of the range the contribution from each data point is the same, but the contribution from the first and last few data points are not equal.

$$\text{Area (parabolic)} = \tfrac{3}{8}(y_1 + y_N) + \tfrac{7}{6}(y_2 + y_{N-1})$$
$$+ \tfrac{23}{24}(y_3 + y_{N-2}) + \sum_{i=4}^{N-3} y_i$$

Note that the sum of the contributions from the six end-data points has the weight corresponding to an average over five data points. This follows from the previous discussion that there are only $N - 1$ intervals between N data points.

Higher-degree polynomials Note that in choosing the intervals for the parabolic interpolation, we were careful to choose them symmetrically placed with respect to the data points for each interpolation except for the end intervals. For a cubic polynomial, we need to choose the intervals between the data points, as shown in Figure 13-4, as they were for a linear interpolation.

Following the same development for a cubic polynomial,

$$y(x) = a + bx + cx^2 + dx^3$$

the area of the first interval has a different functional form from that of the other intervals.

$$A_1 = \tfrac{1}{24}(9y_1 + 19y_2 - 5y_3 + y_4)$$
$$A_i = \tfrac{1}{24}(-y_{i-1} + 13y_i + 13y_{i+1} - y_{i+2}) \qquad i = 2, N - 2$$

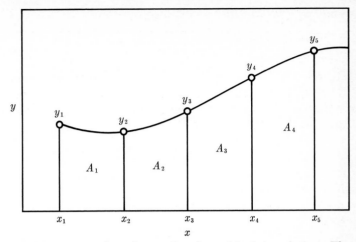

FIGURE 13-4 Area integration by cubic interpolation. The intervals are chosen symmetrically between the data points including the first interval.

The total area contains unequal contributions from eight of the intervals

$$\text{Area (cubic)} = \tfrac{1}{3}(y_1 + y_N) + \tfrac{31}{24}(y_2 + y_{N-1})$$
$$+ \tfrac{5}{6}(y_3 + y_{N-2}) + \tfrac{25}{24}(y_4 + y_{N-3}) + \sum_{i=5}^{N-4} y_i$$

with the sum of these contributions equal to an average over only seven intervals. Note that in all cases the first and last data points are given considerably smaller weights than any of the other data points, and for higher-degree interpolation than linear, the next intervals give the highest contribution of any.

Unequally spaced data points If the data points are not equally spaced in the independent variable x, the preceding formulas are not valid. The same method can be applied but the computation is more elaborate. The procedure is to fit n contiguous data points with an $(n-1)$th-degree polynomial and integrate the fitted curve over one interval in the middle of the fit.

Program 13-3 AREA   Numerical integration of area.

```
C       FUNCTION AREA
C
C       PURPOSE
C          INTEGRATE THE AREA BENEATH A SET OF DATA POINTS
C
C       USAGE
C          RESULT = AREA (X, Y, NPTS, NTERMS)
C
C       DESCRIPTION OF PARAMETERS
C          X       - ARRAY OF DATA POINTS FOR INDEPENDENT VARIABLE
C          Y       - ARRAY OF DATA POINTS FOR DEPENDENT VARIABLE
C          NPTS    - NUMBER OF PAIRS OF DATA POINTS
C          NTERMS  - NUMBER OF TERMS IN FITTING POLYNOMIAL
C
C       SUBROUTINES AND FUNCTION SUBPROGRAMS REQUIRED
C          INTEG (X, Y, NTERMS, I1, X1, X2, SUM)
C             FITS A POLYNOMIAL WITH NTERMS STARTING AT I1
C             AND INTEGRATES AREA FROM X1 TO X2
C
C       MODIFICATIONS FOR FORTRAN II
C          OMIT DOUBLE PRECISION SPECIFICATIONS
C
```

If n is even, the interval is chosen between two data points as in Figure 13-4; if n is odd, the interval straddles one data point as in Figure 13-3, from halfway between one pair of data points to halfway between the next pair. By moving the fit along successive sets of n data points, the entire area is integrated. Contributions from the ends cannot be integrated symmetrically; for these, an $(n - 1)$th-degree polynomial fit is made to the n end points.

Program 13-3   The methods of area integration discussed above are illustrated in the computer routine AREA of Program 13-3. This is a Fortran function subprogram to determine the area beneath a set of data points by fitting successive sets of n contiguous data points with a power-series polynomial. The input variables are X, Y, NPTS, and NTERMS. The data points are assumed to be stored in the arrays X and Y with NPTS $= N$ data points. The number of data points included in each fit is NTERMS $= n$.

Contributions from the ends and the first and last symmetrical intervals are computed in statements 31–39 (for n even) or 51–59 (for n odd). The rest of the intervals are iterated in statements 41–46 or 61–66. If $N \leq n$, the points are fitted with a

Program 13-3 AREA *(continued)*

```
      FUNCTION AREA (X, Y, NPTS, NTERMS)
      DOUBLE PRECISION SUM
      DIMENSION X(1), Y(1)
   11 SUM = 0.
      IF (NPTS - NTERMS) 21, 21, 13
   13 NEVEN = 2*(NTERMS/2)
      IDELTA = NTERMS/2 - 1
      IF (NTERMS - NEVEN) 31, 31, 51
C
C         FIT ALL POINTS WITH ONE CURVE
C
   21 X1 = X(1)
      X2 = X(NPTS)
   23 CALL INTEG (X, Y, NPTS, 1, X1, X2, SUM)
      GO TO 71
C
C         EVEN NUMBER OF TERMS
C
   31 X1 = X(1)
      J = NTERMS - IDELTA
      X2 = X(J)
      CALL INTEG (X, Y, NTERMS, 1, X1, X2, SUM)
      I1 = NPTS - NTERMS + 1
      J = I1 + IDELTA
      X1 = X(J)
      X2 = X(NPTS)
   39 CALL INTEG (X, Y, NTERMS, I1, X1, X2, SUM)
      IF (I1 - 2) 71, 71, 41
   41 IMAX = I1 - 1
      DO 46 I=2, IMAX
      J = I + IDELTA
      X1 = X(J)
      X2 = X(J+1)
   46 CALL INTEG (X, Y, NTERMS, I, X1, X2, SUM)
      GO TO 71
C
C         ODD NUMBER OF TERMS
C
   51 X1 = X(1)
      J = NTERMS - IDELTA
      X2 = (X(J) + X(J-1))/2.
      CALL INTEG (X, Y, NTERMS, 1, X1, X2, SUM)
      I1 = NPTS - NTERMS + 1
      J = I1 + IDELTA
      X1 = (X(J) + X(J+1))/2.
      X2 = X(NPTS)
   59 CALL INTEG (X, Y, NTERMS, I1, X1, X2, SUM)
      IF (I1 - 2) 71, 71, 61
   61 IMAX = I1 - 1
      DO 66 I=2, IMAX
      J = I + IDELTA
      X1 = (X(J+1) + X(J))/2.
      X2 = (X(J+2) + X(J+1))/2.
   66 CALL INTEG (X, Y, NTERMS, I, X1, X2, SUM)
   71 AREA = SUM
      RETURN
      END
```

Program 13-4 INTEG Integration subroutine for AREA.

```
C         SUBROUTINE INTEG
C
C         PURPOSE
C           INTEGRATE THE AREA BENEATH TWO DATA POINTS
C
C         USAGE
C           CALL INTEG (X, Y, NTERMS, I1, X1, X2, SUM)
C
C         DESCRIPTION OF PARAMETERS
C           X      - ARRAY OF DATA POINTS FOR INDEPENDENT VARIABLE
C           Y      - ARRAY OF DATA POINTS FOR DEPENDENT VARIABLE
C           NTERMS - NUMBER OF TERMS IN FITTING POLYNOMIAL
C           I1     - FIRST DATA POINT FOR FITTING POLYNOMIAL
C           X1     - FIRST VALUE OF X FOR INTEGRATION
C           X2     - FINAL VALUE OF X FOR INTEGRATION
C
C         SUBROUTINES AND FUNCTION SUBPROGRAMS REQUIRED
C           NONE
C
C         MODIFICATIONS FOR FORTRAN II
C           OMIT DOUBLE PRECISION SPECIFICATIONS
C
C         COMMENTS
C           DIMENSION STATEMENT VALID FOR NTERMS UP TO 10
C
          SUBROUTINE INTEG (X, Y, NTERMS, I1, X1, X2, SUM)
          DOUBLE PRECISION XJK, ARRAY, A, DENOM, DELTAX, SUM
          DIMENSION X(1), Y(1)
          DIMENSION ARRAY(10, 10)
C
C             CONSTRUCT SQUARE MATRIX AND INVERT
C
       11 DO 17 J=1, NTERMS
          I = J + I1 - 1
          DELTAX = X(I) - X(I1)
          XJK = 1.
          DO 17 K=1, NTERMS
          ARRAY(J,K) = XJK
       17 XJK = XJK * DELTAX
       21 CALL MATINV (ARRAY, NTERMS, DET)
          IF (DET) 31, 23, 31
       23 IMID = I1 + NTERMS/2
          SUM = SUM + Y(IMID)*(X2-X1)
          GO TO 40
C
C             EVALUATE COEFFICIENTS AND INTEGRATE
C
       31 DX1 = X1 - X(I1)
          DX2 = X2 - X(I1)
       33 DO 39 J=1, NTERMS
          I = J + I1 - 1
          A = 0.
          DO 37 K=1, NTERMS
       37 A = A + Y(I)*ARRAY(J,K)
          DENOM = J
       39 SUM = SUM + (A/DENOM)*(DX2**J - DX1**J)
       40 RETURN
          END
```

single curve in statements 21–23. The total area is returned to the main program as the value of the function AREA.

Program 13-4 The computer routine INTEG of Program 13-4 is used to perform the fitting and integration for AREA of Program 13-3. The input variables are X, Y, NTERMS, I1, X1, X2, and SUM, and the output variable is SUM.

The fit to NTERMS $= n$ data points begins with $i = $ I1. The integration is performed between $x = $ X1 and $x = $ X2 which need not correspond to values of the independent data points.

The fitting uses the matrix inversion techniques discussed in Appendix B and used in programs such as LEGFIT, REGRES, CHIFIT, and CURFIT. A square matrix ARRAY, containing values of the n data points to be fitted raised to n powers,

$$\text{ARRAY}(\text{J},\text{K}) = x_i{}^\text{K} \qquad i = \text{J} + \text{I1} - 1$$

is accumulated in statements 11–17 and inverted in statement 21 with the subroutine MATINV of Program B-2. The coefficients of the fit are evaluated in statements 33–37 and used to compute the integrated area in statement 39.

$$\text{Area} = [a_1 x + \tfrac{1}{2}a_2 x^2 + \tfrac{1}{3}a_3 x^3 + \cdot \cdot \cdot]_{x_1 - x_0}^{x_2 - x_0}$$

The variable SUM is increased by the area under the desired interval so that SUM contains the area being accumulated by AREA. Note that the fit considers the first data point $i = $ I1 to be the origin; all values of x are evaluated with respect to $x_0 = $ X(I1).

SUMMARY

Smoothing: For empirical estimates of areas when least-squares fitting is not rigorously valid.

$$y'(x) = \tfrac{1}{4}y(x - 1) + \tfrac{1}{2}y(x) + \tfrac{1}{4}y(x + 1) \qquad \sigma'^2 = \sigma^2 + \tfrac{1}{2}$$

Interpolating between equally spaced variables:

$$y(x) = a_1 + a_2\Delta + a_3\Delta(\Delta - 1) + a_4\Delta(\Delta - 1)(\Delta - 2) + \cdot \cdot \cdot$$

$$a_j = \sum_{i=1}^{j} \left[(-1)^{j-i} \frac{y_i}{(i - 1)!(j - i)!} \right] \qquad \Delta \equiv \frac{x - x_1}{x_2 - x_1}$$

Interpolating between unequally spaced variables:

$$y(x) = a_1 + a_2\Delta + a_3\Delta(\Delta - \Delta_1) + a_4\Delta(\Delta - \Delta_1)(\Delta - \Delta_2) + \cdots$$

$$a_k = \frac{y_k}{\displaystyle\prod_{i=1}^{k-1}(\Delta_k - \Delta_i)} - \sum_{j=1}^{k-1}\frac{a_j}{\displaystyle\prod_{i=j}^{k-1}(\Delta_k - \Delta_i)}$$

$$\Delta \equiv \frac{x - x_1}{x_2 - x_1} \qquad \Delta_i \equiv \frac{x_i - x_1}{x_2 - x_1}$$

Area integration: Fitting n data points exactly with $(n-1)$th-degree polynomial.

Linear: $\text{Area} = \frac{1}{2}(y_1 + y_N) + \displaystyle\sum_{i=2}^{N-1} y_i$

Quadratic: $\text{Area} = \frac{3}{8}(y_1 + y_N) + \frac{7}{6}(y_2 + y_{N-1})$

$$+ \,2\tfrac{3}{24}(y_3 + y_{N-2}) + \sum_{i=4}^{N-3} y_i$$

Cubic: $\text{Area} = \frac{1}{3}(y_1 + y_N) + 3\tfrac{1}{24}(y_2 + y_{N-1})$

$$+ \,\tfrac{5}{6}(y_3 + y_{N-2}) + 2\tfrac{5}{24}(y_4 + y_{N-3}) + \sum_{i=5}^{N-4} y_i$$

EXERCISES

13-1 Interpolate between the data points of Example 8-2 with polynomials of various degrees to find the expected value of y for $\theta = 120°$. Compare with the value predicted by the fitted Legendre polynomial.

13-2 Smooth the data of Example 6-1 and compare with the fitted curve.

13-3 Integrate the area under the data of Example 8-2. Compare with the value computed with the fitted polynomial.

13-4 How would you extend the numerical area integration of Exercise 13-3 to include the area beyond the end data points?

CALCULUS

A-1 PARTIAL DIFFERENTIALS

The derivative of a function is defined as the way in which that function varies for infinitesimal variations in an argument of the function. Let $f(x)$ be a function of some variable x. If x is changed by an amount Δx, the function changes by an amount $\Delta f = f(x + \Delta x) - f(x)$. The ratio $\Delta f/\Delta x$ is a measure of the relative variation of $f(x)$ with x.

In the limit as the change Δx in x is made infinitesimally small, the ratio $\Delta f/\Delta x$ will approach an asymptotic value if the function $f(x)$ is continuous. This asymptotic value is defined to be the *derivative* df/dx of the function $f(x)$ with respect to x.

$$\frac{df(x)}{dx} \equiv \lim_{\Delta x \to 0} \frac{f(x + \Delta x) - f(x)}{\Delta x} \tag{A-1}$$

EXAMPLE A-1   If $f(x) = x^n$, the derivative can be determined by expanding the changed function $f(x + \Delta x)$.

$$\frac{dx^n}{dx} = \lim_{\Delta x \to 0} \frac{x^n + nx^{n-1}\,\Delta x + \Delta x^2[n(n-1)x^{n-2} + \cdots] - x^n}{\Delta x}$$
$$= nx^{n-1} + \lim_{\Delta x \to 0} \{\Delta x[n(n-1)x^{n-2} + \cdots]\} = nx^{n-1}$$

For $n = 4$, for example, $f(x) = x^4$ and $df/dx = 4x^3$.

EXAMPLE A-2   If $f(x) = \sin x$, the sine of a sum (of x and Δx) can be considered as a sum of sines and cosines (of x and Δx).

$$\frac{d(\sin x)}{dx} = \lim_{\Delta x \to 0} \frac{(\sin x)(\cos \Delta x) + (\cos x)(\sin \Delta x) - \sin x}{\Delta x}$$
$$= \frac{\sin x + \Delta x(\cos x) - \sin x}{\Delta x} = \cos x$$

Similarly, if $f(x) = \cos x$, $df/dx = -\sin x$.

Addition and multiplication   The derivative of a sum of functions is equal to the sum of the derivatives. If

$$f(x) = g(x) + h(x)$$

the derivative is also a sum.

$$\frac{df(x)}{dx} = \lim_{\Delta x \to 0} \frac{g(x + \Delta x) - g(x) + h(x + \Delta x) - h(x)}{\Delta x}$$
$$= \frac{dg(x)}{dx} + \frac{dh(x)}{dx}$$

The derivative of a product of functions, however, is not equal to the product of the derivatives. In order to determine what the derivative of a product of functions is, we have to reverse the definition of Equation (A-1).

$$\lim_{\Delta x \to 0} f(x + \Delta x) = \lim_{\Delta x \to 0} \left[f(x) + \Delta x \frac{df(x)}{dx} \right] \tag{A-2}$$

If $f(x) = g(x)h(x)$, the derivative of $f(x)$ is a sum of products.

$$\frac{df(x)}{dx} = \lim_{\Delta x \to 0} \frac{g(x + \Delta x)h(x + \Delta x) - g(x)h(x)}{\Delta x}$$

$$= \lim_{\Delta x \to 0} \frac{1}{\Delta x} \left\{ \left[g(x) + \Delta x \frac{dg(x)}{dx} \right] \left[h(x) + \Delta x \frac{dh(x)}{dx} \right] \right.$$
$$\left. - g(x)h(x) \right\}$$

$$= g(x) \frac{dh(x)}{dx} + h(x) \frac{dg(x)}{dx}$$

Functions of functions If the function $f(x)$ can be expressed as a function of a function $g(x)$ of x,

$$f(x) = f[g(x)]$$

the derivative of $f(x)$ with respect to x can be expressed in terms of the derivative of $g(x)$ with respect to x. If we expand the definition of Equation (A-1) for the derivative,

$$\frac{df(x)}{dx} = \lim_{\Delta x \to 0} \frac{f[g(x + \Delta x)] - f[g(x)]}{\Delta x}$$

we can make use of the relationship of Equation (A-2) to expand still further.

$$\frac{df(x)}{dx} = \lim_{\Delta x \to 0} \frac{f\left[g(x) + \Delta x \frac{dg(x)}{dx} \right] - f[g(x)]}{\Delta x}$$

$$= \lim_{\Delta x \to 0} \frac{f[g(x)] + \Delta x \frac{dg(x)}{dx} \frac{df(x)}{dg(x)} - f[g(x)]}{\Delta x}$$

$$= \frac{df(x)}{dg(x)} \frac{dg(x)}{dx} \qquad\qquad\text{(A-3)}$$

EXAMPLE A-3 If $f(x) = (a - bx^3)^2$, define $g(x) \equiv a + bx^3$ so that $f(x) = [g(x)]^2$. The first term in Equation (A-3) is the derivative of a square, and the second term is the derivative of a cubic polynomial.

$$\frac{df(x)}{dg(x)} = 2g(x) = 2(a + bx^3) \qquad \frac{dg(x)}{dx} = 3bx^2$$

$$\frac{df(x)}{dx} = 2(a + bx^3)3bx^2 = 6bx^2(a + bx^3)$$

Higher-order derivatives Higher-order derivatives are defined as derivatives of derivatives. For example, the second derivative of a function $f(x)$ is just the derivative of the first derivative.

$$\frac{d^2f(x)}{dx^2} \equiv \frac{d}{dx}\left[\frac{df(x)}{dx}\right]$$

For the nth-order derivative $d^nf(x)/dx^n$, we simply take the derivative n times in succession.

For example, if $f(x) = x^4$ as in Example A-1, the second derivative is $12x^2$. Similarly, the fourth derivative of $\sin x$ or of $\cos x$ is equal to itself and equal to the negative of the second derivative.

Partial derivatives If the function $f(x,y)$ is dependent on two variables x and y, we must define derivatives of the function with respect to each of the independent variables. The *partial derivatives* $\partial f/\partial x$ and $\partial f/\partial y$ are defined in the same way as ordinary derivatives, considering all other variables to be constant.

$$\frac{\partial f(x,y)}{\partial x} \equiv \lim_{\Delta x \to 0} \frac{f(x + \Delta x, y) - f(x,y)}{\Delta x} = \frac{df(x)}{dx}$$

$$\frac{\partial f(x,y)}{\partial y} \equiv \lim_{\Delta y \to 0} \frac{f(x, y + \Delta y) - f(x,y)}{\Delta y} = \frac{df(y)}{dy} \qquad \text{(A-4)}$$

Higher-order partial derivatives include not only higher-order derivatives with respect to one variable, but also cross partial derivatives with respect to two or more variables simultaneously.

$$\frac{\partial^2 f(x,y)}{\partial x^2} \equiv \frac{\partial}{\partial x}\left[\frac{\partial f(x,y)}{\partial x}\right]$$

$$\frac{\partial^2 f(x,y)}{\partial x\,\partial y} \equiv \frac{\partial}{\partial x}\left[\frac{\partial f(x,y)}{\partial y}\right] = \frac{\partial}{\partial y}\left[\frac{\partial f(x,y)}{\partial x}\right] = \frac{\partial^2 f(x,y)}{\partial y\,\partial x}$$

In terms of infinitesimal differentials Δx and Δy, we can apply the definitions of Equations (A-4) twice in succession.

$$\frac{\partial^2 f(x,y)}{\partial x\,\partial y} = \lim_{\substack{\Delta x \to 0 \\ \Delta y \to 0}} \frac{f(x + \Delta x, y + \Delta y) - f(x + \Delta x, y) - f(x, y + \Delta y) + f(x,y)}{\Delta x\,\Delta y}$$

$$\frac{\partial^2 f(x,y)}{\partial x^2} = \lim_{\substack{\Delta' x \to 0 \\ \Delta x \to 0}} \frac{f(x + \Delta' x + \Delta x, y) - f(x + \Delta x, y) - f(x + \Delta' x, y) + f(x,y)}{\Delta' x\,\Delta x}$$

The second definition can be made more symmetric by defining $\Delta' x = -\Delta x$ and letting them both go to 0 simultaneously.

$$\frac{\partial^2 f(x,y)}{\partial x^2} = \lim_{\Delta x \to 0} \frac{f(x + \Delta x, y) - 2f(x,y) + f(x - \Delta x, y)}{\Delta x^2}$$

EXAMPLE A-4   If $f(x,y) = ax^2 + bxy^3$, what are the various nonzero partial derivatives?

$$\frac{\partial f}{\partial x} = 2ax + by^3 \qquad \frac{\partial^2 f}{\partial x^2} = 2a$$

$$\frac{\partial f}{\partial y} = 3bxy^2 \qquad \frac{\partial^2 f}{\partial y^2} = 6bxy \qquad \frac{\partial^3 f}{\partial y^3} = 6bx$$

$$\frac{\partial^2 f}{\partial x\,\partial y} = 3by^2 \qquad \frac{\partial^3 f}{\partial x\,\partial y^2} = 6by \qquad \frac{\partial^4 f}{\partial x\,\partial y^3} = 6b$$

A-2   MINIMIZING FUNCTIONS

A function $f(x)$ is said to have a *relative minimum* at the point $x = x_{\min}$ if the value of the function $f(x \pm \Delta x)$ is larger for values of x infinitesimally larger or smaller.

$$f(x_{\min}) < f(x_{\min} \pm \Delta x) \qquad \text{for infinitesimal } \Delta x$$

Similarly, the function $f(x)$ has a *relative maximum* if the function is smaller for values of x infinitesimally larger or smaller. If the value of the function is smaller for $x = x_{\min}$ (or larger for $x = x_{\max}$) than for any other value of x, then the function has an *absolute minimum* (maximum) at $x = x_{\min}(x_{\max})$.

The function illustrated in Figure A-1, for example, has one

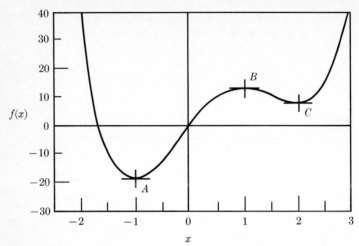

FIGURE A-1 Graph of the function $f(x) = 3x^4 - 8x^3 - 6x^2 + 24x$ vs. x. The absolute minimum occurs for $x = -1$ at A; another relative minimum occurs for $x = 2$ at C; a relative maximum occurs for $x = 1$ at B.

relative maximum at point B and two relative minima, of which the one at point A is an absolute minimum. Note that at the minima and maxima, the slope of the curve is horizontal. Analytically, this is equivalent to noting that the derivative of the function, evaluated at the minima or maxima, is 0.

$$\frac{df(x_{\min})}{dx} \equiv \frac{df(x)}{dx}\bigg|_{x_{\min}} = 0 \tag{A-5}$$

Equation (A-5) provides a convenient method for determining the values of the argument x for which the function $f(x)$ has a relative maximum or minimum: set the derivative of the function with respect to the argument equal to 0 and solve for the roots of the resulting equation. The values of x which satisfy Equation (A-5) are the appropriate solutions.

The question of whether the function has a minimum or a maximum can be resolved by examining the second derivative. If the second derivative is positive, the curvature of the function is upward, as at points A and C of Figure A-1, and the function is at a minimum. If the second derivative is negative, the function has a maximum. If the second derivative is 0 (for example, for $y = x^3$ evaluated at $x = 0$), the function has an inflection which may be a maximum, a minimum, or neither one.

EXAMPLE A-5   The function $f(x) = 3x^4 - 8x^3 - 6x^2 + 24x$ is graphed in Figure A-1. The values of x corresponding to the maxima and minima are the solutions of Equation (A-5).

$$\frac{df(x)}{dx} = 12(x^3 - 2x^2 - x + 2)$$
$$= 12(x + 1)(x - 1)(x - 2) = 0$$

The function has relative minima or maxima at $x = 2$ and $x = \pm 1$. The second derivative

$$\frac{d^2f(x)}{dx^2} = 12(3x^2 - 4x - 1)$$

is positive for $x = 2$ and $x = -1$ and negative for $x = 1$. The latter is therefore the position of a relative maximum, and the other two points correspond to relative minima. The function must be evaluated at the two minima to determine which one is an absolute minimum.

Functions of many variables   If the function $f(x,y)$ is a function of more than one variable, we can still consider the function to have a minimum (or maximum) in parameter space, but we must be careful to define relative minima such that the function has a minimum with respect to all the parameters simultaneously.

The function χ^2, for example,

$$\chi^2 \equiv \sum \left\{ \frac{1}{\sigma_i^2} [y_i - y(x_i)]^2 \right\} = \sum \left(\frac{\Delta y_i}{\sigma_i^2} \right)^2$$

which is discussed in detail in Chapter 10 and used throughout the book, is a function of all the coefficients or parameters a_j of the fitting function $y(x)$.

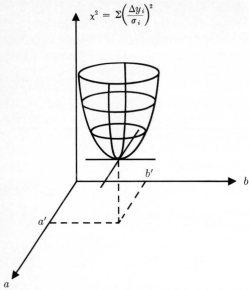

$$\chi^2 = \Sigma\left(\frac{\Delta y_i}{\sigma_i}\right)^2$$

FIGURE A-2 Variation of the weighted sum of squares of the deviations $\chi^2 = \Sigma(\Delta y_i/\sigma_i)^2$ as a function of the coefficients a and b of Equation (6-1). The surface is paraboloid near the vertex where χ^2 has its minimum value.

$$y(x) = \sum_{j=1}^{n} [a_j X_j(x)]$$

Note that χ^2 is not considered to be a function of either x or y but of the parameters a_j. Let us refer to the simple case of a fitting function $y(x) = a + bx$ with two coefficients a and b as discussed in Chapter 6. If we consider the hypersurface in three dimensions describing the variation of χ^2 as a function of a and b, we can construct the paraboloid surface shown in Figure A-2. The function χ^2 is graphed in the vertical direction and the coefficients are given in the horizontal plane.

The smallest value of χ^2 occurs at the bottom vertex of the paraboloid surface, and the corresponding values of $a = a'$ and

$b = b'$ are the appropriate values of the coefficients. At this point the variation of χ^2 for infinitesimally small perturbations of the values of the coefficients is 0. A plane drawn tangent to the surface at this point is horizontal.

As for functions of a single variable, the position of a relative minimum is one for which the derivative of the function is 0, but with the added stipulation that this must be so for the derivative with respect to each of the coefficients (or parameters) simultaneously.

Therefore, we may determine the position of a relative minimum by solving a set of simultaneous equations resulting from setting all the first partial derivatives equal to 0.

$$\frac{\partial}{\partial a} \chi^2 = \frac{\partial}{\partial a} \sum \left[\frac{1}{\sigma_i^2} (y_i - a - bx_i)^2 \right]$$
$$= -2 \sum \left[\frac{1}{\sigma_i^2} (y_i - a - bx_i) \right] = 0$$
$$\frac{\partial}{\partial b} \chi^2 = \frac{\partial}{\partial b} \sum \left[\frac{1}{\sigma_i^2} (y_i - a - bx_i)^2 \right]$$
$$= -2 \sum \left[\frac{x_i}{\sigma_i^2} (y_i - a - bx_i) \right] = 0$$

(A-6)

For a function with n parameters, this procedure results in a set of n simultaneous equations in n unknowns, the solution of which occupies a good portion of the discussion of Chapters 6–11.

The question of whether the solution to Equations (A-6) represents a maximum or minimum may be resolved by examining the second partial derivatives with respect to each of the parameters (but not the cross partial derivatives) to ascertain that they are all positive. In Section 11-3, for example, the curvature of the χ^2 hypersurface is deliberately represented as positive everywhere to force the search toward the point of relative minimum instead of relative maximum.

MATRICES

In applying the method of least squares to both linear and non-linear functions, we have repeatedly required the solution of a set of n simultaneous equations in n unknowns a_i similar to Equations (8-4).

$$\Sigma y_i = a_1\Sigma(1) + a_2\Sigma x_i + a_3\Sigma x_i{}^2$$
$$\Sigma x_i y_i = a_1\Sigma x_i + a_2\Sigma x_i{}^2 + a_3\Sigma x_i{}^3$$
$$\Sigma x_i{}^2 y_i = a_1\Sigma x_i{}^2 + a_2\Sigma x_i{}^3 + a_3\Sigma x_i{}^4$$

The symmetry of the right-hand side suggests that we write elements of the equations in a two-dimensional array

$$\alpha = \begin{bmatrix} \Sigma(1) & \Sigma x_i & \Sigma x_i^2 \\ \Sigma x_i & \Sigma x_i^2 & \Sigma x_i^3 \\ \Sigma x_i^2 & \Sigma x_i^3 & \Sigma x_i^4 \end{bmatrix}$$

and separate the other terms and coefficients into one-dimensional arrays.

$$\beta = \begin{bmatrix} \Sigma y_i \\ \Sigma x_i y_i \\ \Sigma x_i^2 y_i \end{bmatrix} \qquad a = \begin{bmatrix} a_1 & a_2 & a_3 \end{bmatrix}$$

Any such rectangular array is termed a *matrix*. We will be concerned primarily with linear one-dimensional matrices and with symmetric square two-dimensional matrices which have the same number of rows and columns and are mirror symmetric along the diagonal from upper left to lower right. Consider a square matrix **A**.

$$\mathbf{A} = \begin{bmatrix} A_{11} & A_{12} & \cdots & A_{1k} & \cdots & A_{1n} \\ A_{21} & A_{22} & \cdots & A_{2k} & \cdots & A_{2n} \\ \cdots & \cdots & \cdots & \cdots & \cdots & \cdots \\ A_{j1} & A_{j2} & \cdots & A_{jk} & \cdots & A_{jn} \\ \cdots & \cdots & \cdots & \cdots & \cdots & \cdots \\ A_{n1} & A_{n2} & \cdots & A_{nk} & \cdots & A_{nn} \end{bmatrix}$$

The *degree* of the matrix **A** is the number n of rows and columns; the jkth *element* (or *component*) of the matrix is A_{jk}; the *diagonal terms* are A_{jj}. If the matrix is diagonally *symmetric*, $A_{jk} = A_{kj}$, and there are n^2 elements, but only $\frac{1}{2}n(n + 1)$ different elements.

Matrix algebra If **A** and **B** are two square symmetric matrices of degree n, then their sum **S** is a square symmetric matrix of degree n whose elements are the sums of the corresponding elements of the two matrices.

$$\mathbf{A} + \mathbf{B} = \mathbf{S} \qquad S_{jk} = A_{jk} + B_{jk}$$

The product **P** of the matrices **A** and **B** is a square matrix of degree n, but the elements of **P** are determined in a more complex manner; the result may or may not be symmetric.

$$\mathbf{AB} = \mathbf{P} \qquad P_{jk} = \sum_{m=1}^{n} (A_{jm}B_{mk})$$

The elements of the jth row of **A** are multiplied by the elements of the kth column of **B** and the products are summed to obtain the jkth element of **P**.

If a is a linear one-dimensional matrix, the product of **A** and a is only well defined if the product is taken in a particular order. If a is a column matrix, it must be multiplied by the square matrix to yield another column matrix c.

$$\begin{bmatrix} A_{11} & \cdots & \cdots & \cdots & A_{1n} \\ \cdots & \cdots & \cdots & \cdots & \cdots \\ A_{j1} & \cdots & A_{jk} & \cdots & A_{jn} \\ \cdots & \cdots & \cdots & \cdots & \cdots \\ A_{n1} & \cdots & \cdots & \cdots & A_{nn} \end{bmatrix} \begin{bmatrix} a_1 \\ \cdot \\ a_k \\ \cdot \\ a_n \end{bmatrix} = \begin{bmatrix} c_1 \\ \cdot \\ c_j \\ \cdot \\ c_n \end{bmatrix}$$

$$c_j = \sum_{k=1}^{n} (A_{jk}a_k) \quad \text{(B-1)}$$

If a is a row matrix, it must multiply the square matrix to yield another row matrix r.

$$[a_1 \quad \cdots \quad a_j \quad \cdots \quad a_n] \begin{bmatrix} A_{11} & \cdots & A_{1k} & \cdots & A_{1n} \\ \cdots & \cdots & \cdots & \cdots & \cdots \\ \cdots & \cdots & A_{jk} & \cdots & \cdots \\ \cdots & \cdots & \cdots & \cdots & \cdots \\ A_{n1} & \cdots & A_{nk} & \cdots & A_{nn} \end{bmatrix}$$

$$= [r_1 \quad \cdots \quad r_k \quad \cdots \quad r_n] \qquad r_k = \sum_{j=1}^{n} (a_j A_{jk}) \quad \text{(B-2)}$$

The product of two linear matrices depends on the order. The product of a row matrix a times a column matrix b is a scalar. If the order is reversed, the result is a square matrix which is *diagonal*, i.e., for which only the diagonal terms are nonzero.

$$[a_1 \cdots a_n] \begin{bmatrix} b_1 \\ \cdot \\ \cdot \\ \cdot \\ b_n \end{bmatrix} = \sum_{j=1}^{n} (a_j b_j)$$

$$\begin{bmatrix} b_1 \\ \cdot \\ \cdot \\ \cdot \\ b_n \end{bmatrix} [a_1 \cdots a_n] = \begin{bmatrix} a_1 b_1 & & & & \\ & \cdot & & & 0 \\ & & \cdot & & \\ & & & a_j b_j & \\ & & \cdot & & \\ 0 & & & & \cdot \\ & & & & & a_n b_n \end{bmatrix}$$

EXAMPLE B-1 Find the sum $S = A + B$ and the products $P = AB$ and $P' = BA$ of the two square symmetric matrices A and B below.

$$A = \begin{bmatrix} 1 & 2 & 3 \\ 2 & 1 & 1 \\ 3 & 1 & 2 \end{bmatrix} \quad B = \begin{bmatrix} 4 & 5 & 6 \\ 5 & 5 & 6 \\ 6 & 6 & 5 \end{bmatrix}$$

$$S = \begin{bmatrix} 5 & 7 & 9 \\ 7 & 6 & 7 \\ 9 & 7 & 7 \end{bmatrix} \quad P = \begin{bmatrix} 32 & 33 & 33 \\ 19 & 21 & 23 \\ 29 & 32 & 34 \end{bmatrix} \quad P' = \begin{bmatrix} 32 & 19 & 29 \\ 33 & 21 & 32 \\ 33 & 23 & 34 \end{bmatrix}$$

Note the mirror symmetry between P and P'.

EXAMPLE B-2 Find the product of the matrix A of Example B-1 and the linear matrix $a = [4 \quad 5 \quad 6]$ when considered as a row and as a column matrix.

$$[4 \quad 5 \quad 6] \begin{bmatrix} 1 & 2 & 3 \\ 2 & 1 & 1 \\ 3 & 1 & 2 \end{bmatrix} = [32 \quad 19 \quad 29]$$

$$\begin{bmatrix} 1 & 2 & 3 \\ 2 & 1 & 1 \\ 3 & 1 & 2 \end{bmatrix} \begin{bmatrix} 4 \\ 5 \\ 6 \end{bmatrix} = \begin{bmatrix} 32 \\ 19 \\ 29 \end{bmatrix}$$

Determinants The *determinant* of a matrix is defined in terms of its algebra. The *order* of the determinant of a square matrix is equal to the degree n of the matrix. Manipulation of rows may be substituted for that of columns throughout.

1. The determinant of the unity matrix is 1 where the *unity matrix* is defined as the diagonal matrix whose diagonal elements are all equal to 1.

$$|\mathbf{1}| = \begin{vmatrix} 1 & 0 & 0 \\ 0 & 1 & 0 \\ 0 & 0 & 1 \end{vmatrix} = 1$$

2. If a column matrix of degree n is added to one column of a square matrix of degree n, the determinant of the result is the sum of the determinant of the original square matrix plus that of another square matrix obtained by substituting the column matrix for the modified column.

$$\begin{vmatrix} A_{11} + a_1 & A_{12} & A_{13} \\ A_{21} + a_2 & A_{22} & A_{23} \\ A_{31} + a_3 & A_{32} & A_{33} \end{vmatrix} = \begin{vmatrix} A_{11} & A_{12} & A_{13} \\ A_{21} & A_{22} & A_{23} \\ A_{31} & A_{32} & A_{33} \end{vmatrix} + \begin{vmatrix} a_1 & A_{12} & A_{13} \\ a_2 & A_{22} & A_{23} \\ a_3 & A_{32} & A_{33} \end{vmatrix}$$

3. If one column of a square matrix is multiplied by a scalar, the determinant of the result is the product of the scalar and the determinant of the original matrix.

$$\begin{vmatrix} cA_{11} & A_{12} & A_{13} \\ cA_{21} & A_{22} & A_{23} \\ cA_{31} & A_{32} & A_{33} \end{vmatrix} = c \begin{vmatrix} A_{11} & A_{12} & A_{13} \\ A_{21} & A_{22} & A_{23} \\ A_{31} & A_{32} & A_{33} \end{vmatrix}$$

4. If two columns of a square matrix are interchanged, the determinant retains the same magnitude but changes sign.

$$\begin{vmatrix} A_{12} & A_{11} & A_{13} \\ A_{22} & A_{21} & A_{23} \\ A_{32} & A_{31} & A_{33} \end{vmatrix} = - \begin{vmatrix} A_{11} & A_{12} & A_{13} \\ A_{21} & A_{22} & A_{23} \\ A_{31} & A_{32} & A_{33} \end{vmatrix}$$

The *minor* A^{jk} of an element A_{jk} of a square matrix of degree n is defined as the determinant of the square matrix of degree $n - 1$ formed by removing the jth row and the kth column.

$$\mathbf{A} = \begin{bmatrix} A_{11} & A_{12} & A_{13} \\ A_{21} & A_{22} & A_{23} \\ A_{31} & A_{32} & A_{33} \end{bmatrix} \qquad A^{21} = \begin{vmatrix} A_{12} & A_{13} \\ A_{32} & A_{33} \end{vmatrix}$$

The *cofactor* cof (A_{jk}) of an element A_{jk} of a square matrix of degree n is defined as the product of the minor and a phase factor.

$$\text{cof } (A_{jk}) \equiv (-1)^{j+k} A^{jk}$$

With these definitions, the determinant of a square matrix of degree n can be expressed in terms of cofactors or minors.

$$|\mathbf{A}| = \sum_{k=1}^{n} [A_{jk} \text{ cof } (A_{jk})] = \sum_{k=1}^{n} [(-1)^{j+k} A_{jk} A^{jk}] \qquad \text{(B-3)}$$

Equation (B-3) is an iterative definition since the cofactor is itself a determinant. The determinant of a matrix of degree 1, however, is equal to the single element of that matrix. The determinant of a square matrix of degree 2 is encountered often enough to make its explicit formula useful.

$$\begin{vmatrix} a & b \\ c & d \end{vmatrix} = ad - bc$$

Computation For computational purposes, it is simpler for the general case if we can manipulate the matrix to form a diagonal matrix in which only the diagonal elements A_{jj} are nonzero. The determinant of a diagonal matrix is equal to the product of all of the diagonal elements (the *trace* is their sum).

$$|\mathbf{A}_{\text{dias}}| = \prod_{j=1}^{n} A_{jj}$$

If we combine rules (2), (3), and (4) of the algebra for determinants, we can show that the determinant of a matrix is unchanged if the elements of any column, multiplied by an arbitrary scalar, are added to the elements of any other column. The determinant of the sum is equal to the sum of two determinants, but one of these determinants has two identical columns except for a scalar factor which may be extracted.

$$\begin{vmatrix} A_{11} + cA_{12} & A_{12} & A_{13} \\ A_{21} + cA_{22} & A_{22} & A_{23} \\ A_{31} + cA_{32} & A_{32} & A_{33} \end{vmatrix} = \begin{vmatrix} A_{11} & A_{12} & A_{13} \\ A_{21} & A_{22} & A_{23} \\ A_{31} & A_{32} & A_{33} \end{vmatrix}$$

$$+ c \begin{vmatrix} A_{12} & A_{12} & A_{13} \\ A_{22} & A_{22} & A_{23} \\ A_{32} & A_{32} & A_{33} \end{vmatrix} = |\mathbf{A}|$$

By successively subtracting one column or row from each of the others, with the proper normalization, it is possible to eliminate all the elements of one row except for one element. For example, if we perform the following subtraction on each row except the first,

$$A'_{jk} = A_{jk} - A_{1k} \frac{A_{j1}}{A_{11}} \qquad \begin{vmatrix} A_{11} & A_{12} & A_{13} \\ 0 & A'_{22} & A'_{23} \\ 0 & A'_{32} & A'_{33} \end{vmatrix} \qquad \text{(B-4)}$$

all the elements of the first column will vanish except A_{11}.

Similarly, if we subsequently start with element A'_{22} and subtract an appropriately normalized second row from the rest of the rows,

$$A''_{jk} = A'_{jk} - A'_{2k} \frac{A'_{j2}}{A'_{22}} \qquad \begin{vmatrix} A_{11} & 0 & A''_{13} \\ 0 & A'_{22} & A'_{23} \\ 0 & 0 & A''_{33} \end{vmatrix}$$

all the elements of the second column except A'_{22} will vanish. Note, however, that this value of A'_{22} is not the original value of A_{22} but is modified by the first subtraction of row 1.

By successively subtracting rows normalized to their diagonal elements, we can produce a matrix which is diagonal.

In practice it is sufficient to eliminate only half the nondiagonal elements; all the elements on one side of the diagonal must be 0.

$$
\begin{vmatrix} A_{11} & 0 & 0 \\ A_{21} & A_{22} & 0 \\ A_{31} & A_{32} & A_{33} \end{vmatrix} = \begin{vmatrix} A_{11} & A_{12} & A_{13} \\ 0 & A_{22} & A_{23} \\ 0 & 0 & A_{33} \end{vmatrix}
$$

$$
= \begin{vmatrix} A_{11} & 0 & 0 \\ 0 & A_{22} & 0 \\ 0 & 0 & A_{33} \end{vmatrix} = A_{11}A_{22}A_{33}
$$

Program B-1 The computation of the determinant of a matrix is illustrated in the computer routine DETERM of Program B-1. This is a Fortran function subprogram to evaluate the determinant of a square matrix, diagonalizing the matrix in the process. The input variables are ARRAY and NORDER. ARRAY is the matrix whose determinant is to be evaluated, and NORDER $= n$ is the order of the determinant.

The DO loop extending over statements 11–50 iterates through the n diagonal elements. If any diagonal element is 0 when it is encountered, the program interchanges columns to find a diagonal element for that value of K which is not 0, changing the sign of the determinant for the interchange. If no such non-zero element exists, the determinant is 0.

As each diagonal element is encountered, the value of the determinant is multiplied by that value in statement 41, and the DO loop of statements 43–46 serves to eliminate nondiagonal elements as discussed above. When all n diagonal elements have contributed to the value of the determinant, the result is returned to the calling program as the value of the function DETERM.

See also the discussion in Section B-3 of the subroutine MATINV of Program B-2 which evaluates the determinant as a byproduct of matrix inversion.

B-2 SIMULTANEOUS EQUATIONS

A straightforward method of obtaining a solution to a set of n simultaneous equations in n unknowns is to combine the equa-

Program B-1 DETERM Determinant of square matrix.

```
C       FUNCTION DETERM
C
C       PURPOSE
C         CALCULATE THE DETERMINANT OF A SQUARE MATRIX
C
C       USAGE
C         DET = DETERM (ARRAY, NORDER)
C
C       DESCRIPTION OF PARAMETERS
C         ARRAY  - MATRIX
C         NORDER - ORDER OF DETERMINANT (DEGREE OF MATRIX)
C
C       SUBROUTINES AND FUNCTION SUBPROGRAMS REQUIRED
C         NONE
C
C       MODIFICATIONS FOR FORTRAN II
C         OMIT DOUBLE PRECISION SPECIFICATIONS
C
C       COMMENTS
C         THIS SUBPROGRAM DESTROYS THE INPUT MATRIX ARRAY
C         DIMENSION STATEMENT VALID FOR NORDER UP TO 10
C
        FUNCTION DETERM (ARRAY, NORDER)
        DOUBLE PRECISION ARRAY, SAVE
        DIMENSION ARRAY(10,10)
     10 DETERM = 1.
     11 DO 50 K=1, NORDER
C
C          INTERCHANGE COLUMNS IF DIAGONAL ELEMENT IS ZERO
C
        IF (ARRAY(K,K)) 41, 21, 41
     21 DO 23 J=K, NORDER
        IF (ARRAY(K,J)) 31, 23, 31
     23 CONTINUE
        DETERM = 0.
        GO TO 60
     31 DO 34 I=K, NORDER
        SAVE = ARRAY(I,J)
        ARRAY(I,J) = ARRAY(I,K)
     34 ARRAY(I,K) = SAVE
        DETERM = - DETERM
C
C          SUBTRACT ROW K FROM LOWER ROWS TO GET DIAGONAL MATRIX
C
     41 DETERM = DETERM * ARRAY(K,K)
        IF (K - NORDER) 43, 50, 50
     43 K1 = K + 1
        DO 46 I=K1, NORDER
        DO 46 J=K1, NORDER
     46 ARRAY(I,J) = ARRAY(I,J) - ARRAY(I,K)*ARRAY(K,J)/ARRAY(K,K)
     50 CONTINUE
     60 RETURN
        END
```

tions by multiplications and additions to eliminate all but one of the unknowns. Consider, for example, the following set of three equations in three coefficients a_1, a_2, and a_3. We will consider the y_k and X_{jk} to be constants.

$$
\begin{aligned}
y_1 &= a_1 X_{11} + a_2 X_{12} + a_3 X_{13} \\
y_2 &= a_1 X_{21} + a_2 X_{22} + a_3 X_{23} \\
y_3 &= a_1 X_{31} + a_2 X_{32} + a_3 X_{33}
\end{aligned}
\tag{B-5}
$$

We would like to multiply each of the equations by some factor such that when they are added together the constants multiplying a_2 and a_3 are 0.

Determinants   Let us consider the set of three equations as if they were one matrix equation as in Equation (B-1),

$$
\begin{bmatrix} y_1 \\ y_2 \\ y_3 \end{bmatrix} =
\begin{bmatrix} X_{11} & X_{12} & X_{13} \\ X_{21} & X_{22} & X_{23} \\ X_{31} & X_{32} & X_{33} \end{bmatrix}
\begin{bmatrix} a_1 \\ a_2 \\ a_3 \end{bmatrix}
\tag{B-6}
$$

with a and y represented by linear matrices and $\mathbf{X}$ represented by a square matrix. If we multiply the first equation of Equations (B-5) by the cofactor of X_{11} in the matrix of Equation (B-6), multiply the second equation by the cofactor of X_{21}, and multiply the third by the cofactor of X_{31}, then the sum of the three equations is an equation involving determinants according to Equation (B-3).

$$
\begin{vmatrix} y_1 & X_{12} & X_{13} \\ y_2 & X_{22} & X_{23} \\ y_3 & X_{32} & X_{33} \end{vmatrix} = a_1
\begin{vmatrix} X_{11} & X_{12} & X_{13} \\ X_{21} & X_{22} & X_{23} \\ X_{31} & X_{32} & X_{33} \end{vmatrix}
$$

$$
+ a_2 \begin{vmatrix} X_{12} & X_{12} & X_{13} \\ X_{22} & X_{22} & X_{23} \\ X_{32} & X_{32} & X_{33} \end{vmatrix}
+ a_3 \begin{vmatrix} X_{13} & X_{12} & X_{13} \\ X_{23} & X_{22} & X_{23} \\ X_{33} & X_{32} & X_{33} \end{vmatrix}
\tag{B-7}
$$

The determinants in the two rightmost terms of Equation (B-7) both vanish because they have two columns which are identical. Thus, the solution for the coefficient a_1 is the ratio of two determinants.

$$a_1 = \frac{\begin{vmatrix} y_1 & X_{12} & X_{13} \\ y_2 & X_{22} & X_{23} \\ y_3 & X_{32} & X_{33} \end{vmatrix}}{\begin{vmatrix} X_{11} & X_{12} & X_{13} \\ X_{21} & X_{22} & X_{23} \\ X_{31} & X_{32} & X_{33} \end{vmatrix}}$$

The denominator is the determinant of the square matrix $\mathbf{X}$ of Equation (B-6) and the numerator is the determinant of a matrix which is formed by substituting the column matrix y for the first column of the $\mathbf{X}$ matrix.

Similarly, *Cramer's rule* gives the solution for the jth coefficient a_j of a set of n simultaneous equations as the ratio of two determinants.

$$y_k = \sum_{j=1}^{n} (a_j X_{kj}) \qquad k = 1, n \tag{B-8}$$

$$a_j = \frac{|\mathbf{X}'(j)|}{|\mathbf{X}|}$$

The denominator is the determinant of the $\mathbf{X}$ matrix. The numerator $|\mathbf{X}'(j)|$ is the determinant of the matrix formed by substituting the y matrix for the jth column.

A matrix is *singular* if its determinant is 0. If the $\mathbf{X}$ matrix is singular, there is no solution for Equation (B-8). For example, if two of the n simultaneous equations are identical, except for a scale factor, the $\mathbf{X}$ matrix has two identical rows and therefore has a 0 determinant. There are really only $n - 1$ independent simultaneous equations, however, and therefore no solution for n unknowns.

Matrix equation Let us consider Equation (B-8) as if it were a matrix equation as in Equation (B-6). If the $\mathbf{X}$ matrix is square, we can consider the y and a linear matrices as either column matrices as in Equation (B-1) or row matrices as in Equation (B-2).

$$[y_k] = [a_j][X_{kj}] \tag{B-9}$$

If we could multiply this matrix equation by another matrix $\mathbf{X}'$ such that the right-hand side becomes just the linear matrix a, then we will have our solution for the coefficients a_j directly. The multiplication of matrices is associative; that is,

$$\mathbf{A(BC)} = \mathbf{(AB)C}$$

Therefore, we require a matrix $\mathbf{X}'$ such that if it is multiplied by the matrix $\mathbf{X}$, the result is the unity matrix.

$$[X_{kj}][X'_{kj}] = 1 \tag{B-10}$$

The matrix $\mathbf{X}'$ which satisfies Equation (B-10) is called the *inverse matrix* $\mathbf{X}^{-1}$ of $\mathbf{X}$. Equation (B-9) multiplied from the right by $\mathbf{X}^{-1}$ gives the coefficients a_j explicitly, because any matrix is unchanged when multiplied by the unity matrix.

$$[y_k][X_{kj}]^{-1} = [a_j]\,1 = [a_j] \tag{B-11}$$

We can express Equation (B-11) in more conventional form to give the solution for each of the coefficients a_j.

$$a_j = \sum_{k=1}^{n} (y_k X_{kj}^{-1})$$

Thus, the solution of the n unknowns in n simultaneous equations is reduced to evaluating the elements of the inverse matrix $\mathbf{X}^{-1}$.

B-3 MATRIX INVERSION

The *adjoint* $\mathbf{A}^{\dagger}$ of a matrix $\mathbf{A}$ is defined as the matrix obtained by substituting for each element A_{jk} the cofactor of the transposed element A_{kj}.

$$A_{jk}{}^{\dagger} = \text{cof}\,(A_{kj})$$

For a square symmetric matrix, the transposition makes no difference.

The inverse matrix $\mathbf{A}^{-1}$ defined in Equation (B-10) may be evaluated by dividing the adjoint matrix $\mathbf{A}^{\dagger}$ by the determinant of $\mathbf{A}$.

$$A_{jk}^{-1} = \frac{A_{jk}^{\dagger}}{|\mathbf{A}|} \tag{B-12}$$

To show that this equality holds, multiply both sides of Equation (B-12) by $|\mathbf{A}|$ $\mathbf{A}$.

$$|\mathbf{A}| \; \mathbf{A}\mathbf{A}^{-1} = |\mathbf{A}| \; \mathbf{1} = \mathbf{A}\mathbf{A}^{\dagger} \tag{B-13}$$

Diagonal terms of the matrices in Equation (B-13) are equivalent to the formula of Equation (B-3) for evaluating the determinant.

$$|\mathbf{A}| = \sum_{k=1}^{n} (A_{jk}A_{kj}^{\dagger}) = \sum_{k=1}^{n} [A_{jk} \operatorname{cof} (A_{jk})]$$

Off-diagonal elements can be shown to vanish in the same way as the determinants in Equation (B-7). If the matrix $\mathbf{A}$ is singular (that is, if $|\mathbf{A}| = 0$), the inverse matrix $\mathbf{A}^{-1}$ does not exist and there is no solution to the matrix equation of Equation (B-9).

Gauss-Jordan elimination The formula of Equation (B-12) is generally too cumbersome to be used for computing the inverse of a matrix. Instead, the Gauss-Jordan method of elimination is used to invert a matrix by building up a unity matrix into the inverse while reducing the original matrix to a unity matrix.

Consider the inverse matrix $\mathbf{A}^{-1}$ as the ratio of the unity matrix divided by the original matrix $\mathbf{A}^{-1} = 1/\mathbf{A}$. If we manipulate the numerator and denominator of this ratio in the same manner (multiplying rows or columns by the same constant factor and adding the same rows normalized to the same constants), the ratio remains unchanged. If we perform the proper manipulation, we can change the denominator into the unity matrix; the numerator must then become equal to the inverse matrix $\mathbf{A}^{-1}$.

Let us write the 3×3 matrix **A** and the 3×3 unity matrix side by side and manipulate both to reduce the matrix **A** to the unity matrix. We start by using the formula of Equation (B-4) to eliminate the two off-diagonal elements of the first column.

$$
\begin{bmatrix} A_{11} & A_{12} & A_{13} \\ A_{21} & A_{22} & A_{23} \\ A_{31} & A_{32} & A_{33} \end{bmatrix}
\qquad
\begin{bmatrix} 1 & 0 & 0 \\ 0 & 1 & 0 \\ 0 & 0 & 1 \end{bmatrix}
$$

$$
\begin{bmatrix} A_{11} & A_{12} & A_{13} \\ 0 & A_{22} - A_{12}\dfrac{A_{21}}{A_{11}} & A_{23} - A_{13}\dfrac{A_{21}}{A_{11}} \\ 0 & A_{32} - A_{12}\dfrac{A_{31}}{A_{11}} & A_{33} - A_{13}\dfrac{A_{31}}{A_{11}} \end{bmatrix}
\quad
\begin{bmatrix} 1 & 0 & 0 \\ -\dfrac{A_{21}}{A_{11}} & 1 & 0 \\ -\dfrac{A_{31}}{A_{11}} & 0 & 0 \end{bmatrix}
$$

$$\text{(B-14)}$$

Now divide the first row by A_{11} to get a diagonal element of 1.

$$
\begin{bmatrix} 1 & \dfrac{A_{12}}{A_{11}} & \dfrac{A_{13}}{A_{11}} \\ 0 & A_{22} - A_{12}\dfrac{A_{21}}{A_{11}} & A_{23} - A_{13}\dfrac{A_{21}}{A_{11}} \\ 0 & A_{32} - A_{12}\dfrac{A_{31}}{A_{11}} & A_{33} - A_{13}\dfrac{A_{31}}{A_{11}} \end{bmatrix}
\quad
\begin{bmatrix} \dfrac{1}{A_{11}} & 0 & 0 \\ -\dfrac{A_{21}}{A_{11}} & 1 & 0 \\ -\dfrac{A_{31}}{A_{11}} & 0 & 1 \end{bmatrix}
$$

$$\text{(B-15)}$$

The left matrix now has the proper first column. Let us relabel the matrices **B** (on the left) and **B'** (on the right) and perform the corresponding manipulations to the second column.

$$
\begin{bmatrix} 1 & 0 & B_{13} - B_{23}\dfrac{B_{12}}{B_{22}} \\ 0 & 1 & \dfrac{B_{23}}{B_{22}} \\ 0 & 0 & B_{33} - B_{23}\dfrac{B_{32}}{B_{22}} \end{bmatrix}
\quad
\begin{bmatrix} B'_{11} - B'_{21}\dfrac{B_{12}}{B_{22}} & -\dfrac{B_{12}}{B_{22}} & 0 \\ \dfrac{B'_{21}}{B_{22}} & \dfrac{1}{B_{22}} & 0 \\ B'_{31} - B'_{21}\dfrac{B_{32}}{B_{22}} & -\dfrac{B_{32}}{B_{22}} & 1 \end{bmatrix}
$$

$$\text{(B-16)}$$

After similar manipulation of the third column, the matrix on the left becomes the unity matrix and that on the right, therefore, must be the inverse matrix.

For computational purposes, even this method is somewhat inefficient in that two matrices must be manipulated throughout. Note, however, that at each stage of the reduction, there are only three or n useful columns of information in the two matrices. As each column is eliminated from the left matrix, the corresponding column is accumulated on the right.

Therefore, we can combine the manipulation into the range of a single matrix. Start with the matrix **A** and use the formula of Equation (B-4) as for Equation (B-14), but instead of applying this formula to the first column, divide the first column by $-A_{11}$ to get the first column in the right of Equation (B-14); the diagonal element must be divided twice to become $1/A_{11}$. Divide the rest of the first row by A_{11} to get the composite of the two matrices of Equation (B-15).

$$
\begin{bmatrix}
\dfrac{1}{A_{11}} & \dfrac{A_{12}}{A_{11}} & \dfrac{A_{13}}{A_{11}} \\[2ex]
-\dfrac{A_{21}}{A_{11}} & A_{22} - A_{12}\dfrac{A_{21}}{A_{11}} & A_{23} - A_{13}\dfrac{A_{21}}{A_{11}} \\[2ex]
-\dfrac{A_{31}}{A_{11}} & A_{32} - A_{12}\dfrac{A_{31}}{A_{11}} & A_{33} - A_{13}\dfrac{A_{31}}{A_{11}}
\end{bmatrix}
$$

A corresponding manipulation to the second column yields a matrix whose first two columns are identical to those of the right side of Equation (B-16) while the last column is identical to that of the left side of Equation (B-16). The procedure is philosophically the same except that the inverse matrix is being accumulated in the space vacated by the original matrix.

EXAMPLE B-3 Accumulate the inverse of the matrix **A** of Example B-1 by the two elimination methods described above. The first method is illustrated in the first two columns and the other method is illustrated in the third column.

$$
\begin{bmatrix} 1 & 2 & 3 \\ 2 & 1 & 1 \\ 3 & 1 & 2 \end{bmatrix}
\qquad
\begin{bmatrix} 1 & 0 & 0 \\ 0 & 1 & 0 \\ 0 & 0 & 1 \end{bmatrix}
\qquad
\begin{bmatrix} 1 & 2 & 3 \\ 2 & 1 & 1 \\ 3 & 1 & 2 \end{bmatrix}
$$

$$
\begin{bmatrix} 1 & 2 & 3 \\ 0 & -3 & -5 \\ 0 & -5 & -7 \end{bmatrix}
\qquad
\begin{bmatrix} 1 & 0 & 0 \\ -2 & 1 & 0 \\ -3 & 0 & 1 \end{bmatrix}
\qquad
\begin{bmatrix} 1 & 2 & 3 \\ -2 & -3 & -5 \\ -3 & -5 & -7 \end{bmatrix}
$$

$$
\begin{bmatrix} 1 & 0 & -\tfrac{1}{3} \\ 0 & 1 & \tfrac{5}{3} \\ 0 & 0 & \tfrac{4}{3} \end{bmatrix}
\qquad
\begin{bmatrix} -\tfrac{1}{3} & \tfrac{2}{3} & 0 \\ \tfrac{2}{3} & -\tfrac{1}{3} & 0 \\ \tfrac{1}{3} & -\tfrac{5}{3} & 1 \end{bmatrix}
\qquad
\begin{bmatrix} -\tfrac{1}{3} & \tfrac{2}{3} & -\tfrac{1}{3} \\ \tfrac{2}{3} & -\tfrac{1}{3} & \tfrac{5}{3} \\ \tfrac{1}{3} & -\tfrac{5}{3} & \tfrac{4}{3} \end{bmatrix}
$$

$$
\begin{bmatrix} 1 & 0 & 0 \\ 0 & 1 & 0 \\ 0 & 0 & 1 \end{bmatrix}
\qquad
\begin{bmatrix} -\tfrac{1}{4} & \tfrac{1}{4} & \tfrac{1}{4} \\ \tfrac{1}{4} & \tfrac{7}{4} & -\tfrac{5}{4} \\ \tfrac{1}{4} & -\tfrac{5}{4} & \tfrac{3}{4} \end{bmatrix}
\qquad
\begin{bmatrix} -\tfrac{1}{4} & \tfrac{1}{4} & \tfrac{1}{4} \\ \tfrac{1}{4} & \tfrac{7}{4} & -\tfrac{5}{4} \\ \tfrac{1}{4} & -\tfrac{5}{4} & \tfrac{3}{4} \end{bmatrix}
$$

Program B-2 The computation of matrix inversion is illustrated in the computer routine MATINV[1] of Program B-2. This is a Fortran subroutine to invert a square symmetric matrix and calculate its determinant, substituting the inverse matrix into the same array as the original matrix. The input variables are ARRAY and NORDER, and the output variable is DET. ARRAY is the matrix whose inverse is to be evaluated, NORDER = n is the degree of the matrix (order of its determinant), and DET is the value of the determinant.

The DO loop extending over statements 11–100 iterates through the n columns of the matrix. The matrix is reorganized, if necessary, in statements 21–60 to get the largest element in the diagonal in order to improve the computational precision. The inversion procedure discussed above is carried out in statements 61–100 with the determinant DET accumulated in statement 100 as the trace of the diagonalized matrix. The elements of the appropriate column are evaluated in statements 61–70, and those of the corresponding row are evaluated in statements 81–90. All other elements are accumulated in statements 71–80.

After inversion, the matrix is restored to its original ordering

[1] The subroutine MATINV follows the procedure of the subroutine MINV of the IBM "System/360 Scientific Subroutine Package."

Program B-2   MATINV   Matrix inversion.

```
C       SUBROUTINE MATINV
C
C       PURPOSE
C         INVERT A SYMMETRIC MATRIX AND CALCULATE ITS DETERMINANT
C
C       USAGE
C         CALL MATINV (ARRAY, NORDER, DET)
C
C       DESCRIPTION OF PARAMETERS
C         ARRAY  - INPUT MATRIX WHICH IS REPLACED BY ITS INVERSE
C         NORDER - DEGREE OF MATRIX (ORDER OF DETERMINANT)
C         DET    - DETERMINANT OF INPUT MATRIX
C
C       SUBROUTINES AND FUNCTION SUBPROGRAMS REQUIRED
C         NONE
C
C       MODIFICATIONS FOR FORTRAN II
C         OMIT DOUBLE PRECISION SPECIFICATIONS
C         CHANGE DABS TO ABSF IN STATEMENT 23
C
C       COMMENTS
C         DIMENSION STATEMENT VALID FOR NORDER UP TO 10
C
        SUBROUTINE MATINV (ARRAY, NORDER, DET)
        DOUBLE PRECISION ARRAY, AMAX, SAVE
        DIMENSION ARRAY(10,10), IK(10), JK(10)
   10 DET = 1.
   11 DO 100 K=1, NORDER
C
C           FIND LARGEST ELEMENT ARRAY(I,J) IN REST OF MATRIX
C
        AMAX = 0.
   21 DO 30 I=K, NORDER
        DO 30 J=K, NORDER
   23 IF (DABS(AMAX) - DABS(ARRAY(I,J))) 24, 24, 30
   24 AMAX = ARRAY(I,J)
        IK(K) = I
        JK(K) = J
   30 CONTINUE
C
C           INTERCHANGE ROWS AND COLUMNS TO PUT AMAX IN ARRAY(K,K)
C
   31 IF (AMAX) 41, 32, 41
   32 DET = 0.
        GO TO 140
   41 I = IK(K)
        IF (I-K) 21, 51, 43
   43 DO 50 J=1, NORDER
        SAVE = ARRAY(K,J)
        ARRAY(K,J) = ARRAY(I,J)
   50 ARRAY(I,J) = -SAVE
   51 J = JK(K)
        IF (J-K) 21, 61, 53
   53 DO 60 I=1, NORDER
        SAVE = ARRAY(I,K)
        ARRAY(I,K) = ARRAY(I,J)
   60 ARRAY(I,J) = -SAVE
```

```
C
C          ACCUMULATE ELEMENTS OF INVERSE MATRIX
C
   61 DO 70 I=1, NORDER
      IF (I-K) 63, 70, 63
   63 ARRAY(I,K) = -ARRAY(I,K) / AMAX
   70 CONTINUE
   71 DO 80 I=1, NORDER
      DO 80 J=1, NORDER
      IF (I-K) 74, 80, 74
   74 IF (J-K) 75, 80, 75
   75 ARRAY(I,J) = ARRAY(I,J) + ARRAY(I,K)*ARRAY(K,J)
   80 CONTINUE
   81 DO 90 J=1, NORDER
      IF (J-K) 83, 90, 83
   83 ARRAY(K,J) = ARRAY(K,J) / AMAX
   90 CONTINUE
      ARRAY(K,K) = 1. / AMAX
  100 DET = DET * AMAX
C
C          RESTORE ORDERING OF MATRIX
C
  101 DO 130 L=1, NORDER
      K = NORDER - L + 1
      J = IK(K)
      IF (J-K) 111, 111, 105
  105 DO 110 I=1, NORDER
      SAVE = ARRAY(I,K)
      ARRAY(I,K) = -ARRAY(I,J)
  110 ARRAY(I,J) = SAVE
  111 I = JK(K)
      IF (I-K) 130, 130, 113
  113 DO 120 J=1, NORDER
      SAVE = ARRAY(K,J)
      ARRAY(K,J) = -ARRAY(I,J)
  120 ARRAY(I,J) = SAVE
  130 CONTINUE
  140 RETURN
      END
```

in statements 101–130, corresponding to the reorganization performed above. The inverted matrix is returned to the main calling program in place of the original matrix ARRAY, and the determinant is returned as the value of DET.

GRAPHS AND TABLES

C-1 GAUSSIAN PROBABILITY DISTRIBUTION

The probability function $P_G(x,\mu,\sigma)$ for the Gaussian or normal error distribution is given by

$$P_G(x,\mu,\sigma) = \frac{1}{\sigma\sqrt{2\pi}} \exp\left[-\frac{1}{2}\left(\frac{x-\mu}{\sigma}\right)^2\right]$$

If measurements of a quantity x are distributed in this manner around a mean μ with a standard deviation σ, the probability $dP_G(x,\mu,\sigma)$ for observing a value of x, within an infinitesimally small interval dx, in a random sample measurement is given by

$$dP_G(x,\mu,\sigma) = P_G(x,\mu,\sigma)\, dx$$

Values of the probability function $P_G(x,\mu,\sigma)$ are tabulated in Table C-1 as a function of the dimensionless deviation

$$z = |x - \mu|/\sigma$$

for z ranging from 0.0 to 3.0 in increments of 0.01 and up to 5.9 in increments of 0.1. These values were calculated with the subroutine PGAUSS of Program 3-4. This function is graphed on a semilogarithmic scale as a function of z in Figure C-1.

The function which is tabulated and graphed is $P_G(z,0,1)$ which gives the probability that $x = \mu \pm z\sigma$. It is the curve of Figure 3-5 tabulated only for positive values of z as indicated.

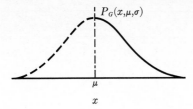

$P_G(x,\mu,\sigma)$

μ

x

Table C-1 Gaussian probability distribution. The Gaussian or normal error distribution $P_G(x,\mu,\sigma)$ vs. $z = |x - \mu|/\sigma$

z	.00	.01	.02	.03	.04	.05	.06	.07	.08	.09
0.0	.39894	.39892	.39886	.39876	.39862	.39844	.39822	.39797	.39767	.39733
0.1	.39695	.39654	.39608	.39559	.39505	.39448	.39387	.39322	.39253	.39181
0.2	.39104	.39024	.38940	.38853	.38762	.38667	.38568	.38466	.38361	.38251
0.3	.38139	.38023	.37903	.37780	.37654	.37524	.37391	.37255	.37115	.36973
0.4	.36827	.36678	.36526	.36371	.36213	.36053	.35889	.35723	.35553	.35381
0.5	.35207	.35029	.34849	.34667	.34482	.34294	.34105	.33912	.33718	.33521
0.6	.33322	.33121	.32918	.32713	.32506	.32297	.32086	.31874	.31659	.31443
0.7	.31225	.31006	.30785	.30563	.30339	.30114	.29887	.29659	.29431	.29200
0.8	.28969	.28737	.28504	.28269	.28034	.27799	.27562	.27324	.27086	.26848
0.9	.26609	.26369	.26129	.25888	.25647	.25406	.25164	.24923	.24681	.24439
1.0	.24197	.23955	.23713	.23471	.23230	.22988	.22747	.22506	.22266	.22025
1.1	.21785	.21546	.21307	.21069	.20831	.20594	.20357	.20122	.19887	.19652
1.2	.19419	.19186	.18955	.18724	.18494	.18265	.18038	.17811	.17585	.17361
1.3	.17137	.16915	.16694	.16475	.16256	.16039	.15823	.15609	.15395	.15184
1.4	.14973	.14764	.14557	.14351	.14147	.13944	.13742	.13543	.13344	.13148
1.5	.12952	.12759	.12567	.12377	.12189	.12002	.11816	.11633	.11451	.11271
1.6	.11093	.10916	.10741	.10568	.10397	.10227	.10059	.09893	.09729	.09567
1.7	.09406	.09247	.09090	.08934	.08780	.08629	.08478	.08330	.08184	.08039
1.8	.07896	.07755	.07615	.07477	.07342	.07207	.07075	.06944	.06815	.06688
1.9	.06562	.06439	.06316	.06196	.06077	.05960	.05845	.05731	.05619	.05509
2.0	.05400	.05293	.05187	.05083	.04981	.04880	.04781	.04683	.04587	.04492
2.1	.04399	.04307	.04217	.04129	.04041	.03956	.03871	.03788	.03707	.03627
2.2	.03548	.03471	.03395	.03320	.03247	.03175	.03104	.03034	.02966	.02899
2.3	.02833	.02769	.02705	.02643	.02582	.02522	.02464	.02406	.02350	.02294
2.4	.02240	.02187	.02135	.02083	.02033	.01984	.01936	.01889	.01843	.01798
2.5	.01753	.01710	.01667	.01626	.01585	.01545	.01506	.01468	.01431	.01394
2.6	.01359	.01324	.01290	.01256	.01224	.01192	.01160	.01130	.01100	.01071
2.7	.01042	.01015	.00987	.00961	.00935	.00910	.00885	.00861	.00837	.00814
2.8	.00792	.00770	.00749	.00728	.00707	.00688	.00668	.00649	.00631	.00613
2.9	.00595	.00578	.00562	.00546	.00530	.00514	.00500	.00485	.00471	.00457

	.00	.10	.20	.30	.40
3.0	.0044318	.0032668	.0023841	.0017226	.0012322
3.5	.00087269	.00061191	.00042479	.00029195	.00019866
4.0	.00013383	.000089264	.000058945	.000038536	.000024943
4.5	.000015984	.000010141	.0000063701	.0000039615	.0000024391
5.0	.0000014868	.00000089730	.00000053614	.00000031716	.00000018575
5.5	.00000010771	.00000006183	.00000003514	.00000001978	.00000001102

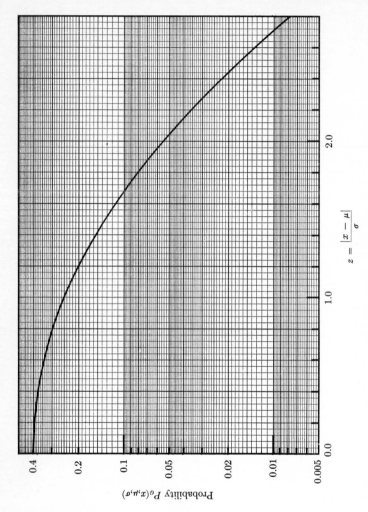

FIGURE C-1 The Gaussian probability function $P_G(x,\mu,\sigma)$ vs. $z = |x - \mu|/\sigma$.

C-2   INTEGRAL OF GAUSSIAN DISTRIBUTION

The integral $A_G(x,\mu,\sigma)$ of the probability function $P_G(x,\mu,\sigma)$ for the Gaussian or normal error distribution is given by

$$A_G(x,\mu,\sigma) = \frac{1}{\sigma\sqrt{2\pi}} \int_{\mu-z\sigma}^{\mu+z\sigma} \exp\left[-\frac{1}{2}\left(\frac{x-\mu}{\sigma}\right)^2\right] dx$$

$$z = \frac{|x-\mu|}{\sigma}$$

If measurements of a quantity x are distributed according to the Gaussian distribution around a mean μ with a standard deviation σ, $A_G(x,\mu,\sigma)$ is equal to the probability for observing a value of x in a random sample measurement which is between $\mu - z\sigma$ and $\mu + z\sigma$; that is, it is the probability that $|x - \mu| < z\sigma$.

Values of the integral $A_G(x,\mu,\sigma)$ are tabulated in Table C-2 as a function of z for z ranging from 0.0 to 3.0 in increments of 0.01 and up to 5.9 in increments of 0.1. These values were calculated with the subroutine AGAUSS of Program 3-5. This function is graphed on a probability scale as a function of z in Figure C-2.

A related function is the error function erf Z.

$$\text{erf } Z = \frac{1}{\sqrt{\pi}} \int_{-Z}^{Z} e^{-z^2}\, dz = A_G(z\sqrt{2}, 0,1)$$

The function which is tabulated and graphed is the shaded area between the limits $\mu \pm z\sigma$ as indicated.

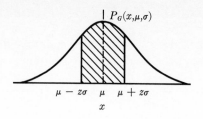

Table C-2 Integral of Gaussian distribution. The integral of the Gaussian probability distribution $A_G(x,\mu,\sigma)$ vs. $z = |x - \mu|/\sigma$

z	.00	.01	.02	.03	.04	.05	.06	.07	.08	.09
0.0	.0	.00798	.01596	.02393	.03191	.03988	.04784	.05581	.06376	.07171
0.1	.07966	.08759	.09552	.10343	.11134	.11924	.12712	.13499	.14285	.15069
0.2	.15852	.16633	.17413	.18191	.18967	.19741	.20514	.21284	.22052	.22818
0.3	.23582	.24344	.25103	.25860	.26614	.27366	.28115	.28862	.29605	.30346
0.4	.31084	.31819	.32551	.33280	.34006	.34729	.35448	.36164	.36877	.37587
0.5	.38292	.38995	.39694	.40389	.41080	.41768	.42452	.43132	.43809	.44481
0.6	.45149	.45814	.46474	.47131	.47783	.48431	.49075	.49714	.50350	.50981
0.7	.51607	.52230	.52847	.53461	.54070	.54674	.55274	.55870	.56461	.57047
0.8	.57629	.58206	.58778	.59346	.59909	.60467	.61021	.61570	.62114	.62653
0.9	.63188	.63718	.64243	.64763	.65278	.65789	.66294	.66795	.67291	.67783
1.0	.68269	.68750	.69227	.69699	.70166	.70628	.71085	.71538	.71985	.72428
1.1	.72866	.73300	.73728	.74152	.74571	.74985	.75395	.75799	.76199	.76595
1.2	.76985	.77371	.77753	.78130	.78502	.78869	.79232	.79591	.79945	.80294
1.3	.80639	.80980	.81316	.81647	.81975	.82298	.82616	.82930	.83240	.83546
1.4	.83848	.84145	.84438	.84727	.85012	.85293	.85570	.85843	.86112	.86377
1.5	.86638	.86895	.87148	.87397	.87643	.87885	.88123	.88358	.88588	.88816
1.6	.89039	.89259	.89476	.89689	.89898	.90105	.90308	.90507	.90703	.90896
1.7	.91086	.91272	.91456	.91636	.91813	.91987	.92158	.92326	.92491	.92654
1.8	.92813	.92969	.93123	.93274	.93422	.93568	.93711	.93851	.93988	.94123
1.9	.94256	.94386	.94513	.94638	.94761	.94882	.95000	.95115	.95229	.95340
2.0	.95449	.95556	.95661	.95764	.95864	.95963	.96059	.96154	.96247	.96338
2.1	.96426	.96513	.96599	.96682	.96764	.96844	.96922	.96999	.97074	.97147
2.2	.97219	.97289	.97358	.97425	.97490	.97555	.97617	.97679	.97739	.97797
2.3	.97855	.97911	.97965	.98019	.98071	.98122	.98172	.98221	.98268	.98315
2.4	.98360	.98404	.98448	.98491	.98531	.98571	.98610	.98648	.98686	.98722
2.5	.98758	.98792	.98826	.98859	.98891	.98922	.98953	.98983	.99012	.99040
2.6	.99067	.99094	.99120	.99146	.99171	.99195	.99218	.99241	.99264	.99285
2.7	.99306	.99327	.99347	.99366	.99385	.99404	.99422	.99439	.99456	.99473
2.8	.99489	.99504	.99520	.99534	.99549	.99563	.99576	.99589	.99602	.99615
2.9	.99627	.99638	.99650	.99661	.99672	.99682	.99692	.99702	.99712	.99721

	.00	.10	.20	.30	.40
3.0	.9973002	.9980648	.9986257	.99903315	.99932614
3.5	.99953474	.99968178	.99978440	.99985530	.999903805
4.0	.999936656	.999958684	.999973308	.999982920	.999989174
4.5	.9999932043	.9999957748	.9999973982	.9999984132	.99999904149
5.0	.99999942657	.99999966024	.99999980061	.99999988410	.99999993327
5.5	.99999996193	.99999997847	.99999998793	.99999999328	.99999999627

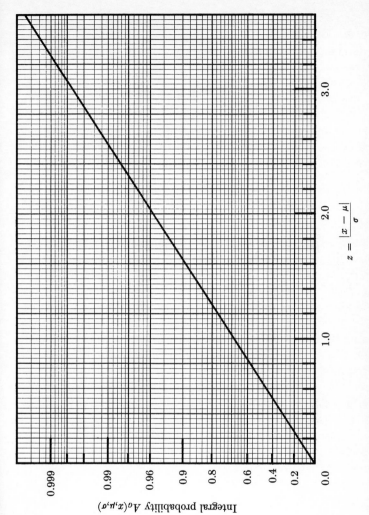

FIGURE C-2 The integral of the Gaussian probability distributions $A_G(x,\mu,\sigma)$ vs. $z = |x - \mu|/\sigma$.

C-3 LINEAR - CORRELATION COEFFICIENT

The probability distribution $P_r(r,\nu)$ for the linear-correlation coefficient r for ν degrees of freedom is given by

$$P_r(r,\nu) = \frac{1}{\sqrt{\pi}} \frac{\Gamma[(\nu + 1)/2]}{\Gamma(\nu/2)} (1 - r^2)^{\frac{1}{2}(\nu-2)}$$

The probability of observing a value of the correlation coefficient larger than r for a random sample of N observations with ν degrees of freedom is the integral of this probability $P_c(r,N)$.

$$P_c(r,N) = \frac{1}{\sqrt{\pi}} \frac{\Gamma[(\nu + 1)/2]}{\Gamma(\nu/2)} \int_{|r|}^{1} (1 - x^2)^{\frac{1}{2}(\nu-2)} \, dx$$

$$\nu = N - 2$$

If two variables of a parent population are uncorrelated, the probability that a random sample of N observations will yield a correlation coefficient for those two variables greater in magnitude than $|r|$ is given by $P_c(r,N)$.

Values of the coefficient $|r|$ corresponding to various values of the probability $P_c(r,N)$ are tabulated in Table C-3 for N ranging from 3 to 100, and values of $P_c(r,N)$ ranging from 0.001 to 0.5. These values were calculated with the subroutine PCORRE of Program 7-1. The functional dependence of r corresponding to representative values of $P_c(r,N)$ is graphed on a semilogarithmic scale as a smooth variation with the number of observations N in Figure C-3.

The function which is tabulated and graphed is the shaded area under the tails of the probability curve for values larger than $|r|$ as indicated.

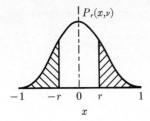

Table C-3 Linear-correlation coefficient. The linear-correlation coefficient r vs. the number of observations N and the corresponding probability $P_c(r,N)$ of exceeding r in a random sample of observations taken from an uncorrelated parent population ($\rho = 0$)

N	P 0.50	0.20	0.10	0.050	0.020	0.010	0.005	0.002	0.001
3	0.707	0.951	0.988	0.997	1.000	1.000	1.000	1.000	1.000
4	0.500	0.800	0.900	0.950	0.980	0.990	0.995	0.998	0.999
5	0.404	0.687	0.805	0.878	0.934	0.959	0.974	0.986	0.991
6	0.347	0.608	0.729	0.811	0.882	0.917	0.942	0.963	0.974
7	0.309	0.551	0.669	0.754	0.833	0.875	0.906	0.935	0.951
8	0.281	0.507	0.621	0.707	0.789	0.834	0.870	0.905	0.925
9	0.260	0.472	0.582	0.666	0.750	0.798	0.836	0.875	0.898
10	0.242	0.443	0.549	0.632	0.715	0.765	0.805	0.847	0.872
11	0.228	0.419	0.521	0.602	0.685	0.735	0.776	0.820	0.847
12	0.216	0.398	0.497	0.576	0.658	0.708	0.750	0.795	0.823
13	0.206	0.380	0.476	0.553	0.634	0.684	0.726	0.772	0.801
14	0.197	0.365	0.458	0.532	0.612	0.661	0.703	0.750	0.780
15	0.189	0.351	0.441	0.514	0.592	0.641	0.683	0.730	0.760
16	0.182	0.338	0.426	0.497	0.574	0.623	0.664	0.711	0.742
17	0.176	0.327	0.412	0.482	0.558	0.606	0.647	0.694	0.725
18	0.170	0.317	0.400	0.468	0.543	0.590	0.631	0.678	0.708
19	0.165	0.308	0.389	0.456	0.529	0.575	0.616	0.662	0.693
20	0.160	0.299	0.378	0.444	0.516	0.561	0.602	0.648	0.679
22	0.152	0.284	0.360	0.423	0.492	0.537	0.576	0.622	0.652
24	0.145	0.271	0.344	0.404	0.472	0.515	0.554	0.599	0.629
26	0.138	0.260	0.330	0.388	0.453	0.496	0.534	0.578	0.607
28	0.133	0.250	0.317	0.374	0.437	0.479	0.515	0.559	0.588
30	0.128	0.241	0.306	0.361	0.423	0.463	0.499	0.541	0.570
32	0.124	0.233	0.296	0.349	0.409	0.449	0.484	0.526	0.554
34	0.120	0.225	0.287	0.339	0.397	0.436	0.470	0.511	0.539
36	0.116	0.219	0.279	0.329	0.386	0.424	0.458	0.498	0.525
38	0.113	0.213	0.271	0.320	0.376	0.413	0.446	0.486	0.513
40	0.110	0.207	0.264	0.312	0.367	0.403	0.435	0.474	0.501
42	0.107	0.202	0.257	0.304	0.358	0.393	0.425	0.463	0.490
44	0.104	0.197	0.251	0.297	0.350	0.384	0.416	0.453	0.479
46	0.102	0.192	0.246	0.291	0.342	0.376	0.407	0.444	0.469
48	0.100	0.188	0.240	0.285	0.335	0.368	0.399	0.435	0.460
50	0.098	0.184	0.235	0.279	0.328	0.361	0.391	0.427	0.451
60	0.089	0.168	0.214	0.254	0.300	0.330	0.358	0.391	0.414
70	0.082	0.155	0.198	0.235	0.278	0.306	0.332	0.363	0.385
80	0.077	0.145	0.185	0.220	0.260	0.286	0.311	0.340	0.361
90	0.072	0.136	0.174	0.207	0.245	0.270	0.293	0.322	0.341
100	0.068	0.129	0.165	0.197	0.232	0.256	0.279	0.305	0.324

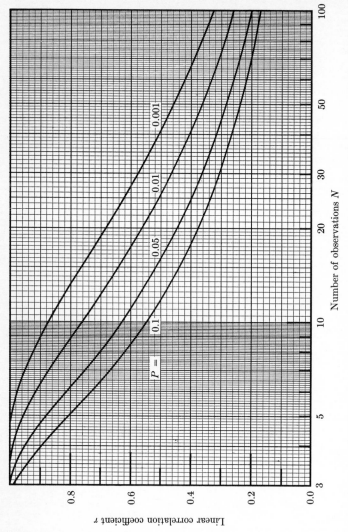

FIGURE C-3 The linear-correlation coefficient r vs. the number of observations N and the corresponding probability $P_c(r,N)$ that the variables are not correlated.

C-4 χ^2 DISTRIBUTION

The probability distribution $P_x(x^2,\nu)$ for χ^2 is given by

$$P_x(x^2,\nu) = \frac{1}{2^{\nu/2}\Gamma(\nu/2)}\,(x^2)^{\frac{1}{2}(\nu-2)}\,e^{-x^2/2}$$

The probability of observing a value of chi-square larger than χ^2 for a random sample of N observations with ν degrees of freedom is the integral of this probability $P_\chi(\chi^2,\nu)$.

$$P_\chi(\chi^2,\nu) = \frac{1}{2^{\nu/2}\Gamma(\nu/2)}\int_{\chi^2}^{\infty}(x^2)^{\frac{1}{2}(\nu-2)}e^{-x^2/2}d(x^2)$$

Values of the reduced chi-square $\chi_\nu^2 = \chi^2/\nu$ corresponding to various values of the integral probability $P_\chi(\chi^2,\nu)$ of exceeding χ^2 in a measurement with ν degrees of freedom are tabulated in Table C-4 for ν ranging from 1 to 200. These values were calculated with the subroutine PCHISQ of Program 10-1. The functional dependence of $P_\chi(\chi^2,\nu)$ corresponding to representative values of ν is graphed in Figure C-4 as a smooth variation with the reduced chi-square χ_ν^2.

The function which is tabulated and graphed is the shaded area under the tail of the probability curve for values larger than χ^2 as indicated.

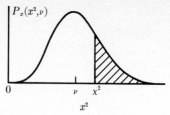

Table C-4 χ^2 distribution. Values of the reduced chi-square $\chi_\nu^2 = \chi^2/\nu$ corresponding to the probability $P_\chi(\chi^2,\nu)$ of exceeding χ^2 vs. the number of degrees of freedom ν

ν \ P	0.99	0.98	0.95	0.90	0.80	0.70	0.60	0.50
1	0.00016	0.00063	0.00393	0.0158	0.0642	0.148	0.275	0.455
2	0.0100	0.0202	0.0515	0.105	0.223	0.357	0.511	0.693
3	0.0383	0.0617	0.117	0.195	0.335	0.475	0.623	0.789
4	0.0742	0.107	0.178	0.266	0.412	0.549	0.688	0.839
5	0.111	0.150	0.229	0.322	0.469	0.600	0.731	0.870
6	0.145	0.189	0.273	0.367	0.512	0.638	0.762	0.891
7	0.177	0.223	0.310	0.405	0.546	0.667	0.785	0.907
8	0.206	0.254	0.342	0.436	0.574	0.691	0.803	0.918
9	0.232	0.281	0.369	0.463	0.598	0.710	0.817	0.927
10	0.256	0.306	0.394	0.487	0.618	0.727	0.830	0.934
11	0.278	0.328	0.416	0.507	0.635	0.741	0.840	0.940
12	0.298	0.348	0.436	0.525	0.651	0.753	0.848	0.945
13	0.316	0.367	0.453	0.542	0.664	0.764	0.856	0.949
14	0.333	0.383	0.469	0.556	0.676	0.773	0.863	0.953
15	0.349	0.399	0.484	0.570	0.687	0.781	0.869	0.956
16	0.363	0.413	0.498	0.582	0.697	0.789	0.874	0.959
17	0.377	0.427	0.510	0.593	0.706	0.796	0.879	0.961
18	0.390	0.439	0.522	0.604	0.714	0.802	0.883	0.963
19	0.402	0.451	0.532	0.613	0.722	0.808	0.887	0.965
20	0.413	0.462	0.543	0.622	0.729	0.813	0.890	0.967
22	0.434	0.482	0.561	0.638	0.742	0.823	0.897	0.970
24	0.452	0.500	0.577	0.652	0.753	0.831	0.902	0.972
26	0.469	0.516	0.592	0.665	0.762	0.838	0.907	0.974
28	0.484	0.530	0.605	0.676	0.771	0.845	0.911	0.976
30	0.498	0.544	0.616	0.687	0.779	0.850	0.915	0.978
32	0.511	0.556	0.627	0.696	0.786	0.855	0.918	0.979
34	0.523	0.567	0.637	0.704	0.792	0.860	0.921	0.980
36	0.534	0.577	0.646	0.712	0.798	0.864	0.924	0.982
38	0.545	0.587	0.655	0.720	0.804	0.868	0.926	0.983
40	0.554	0.596	0.663	0.726	0.809	0.872	0.928	0.983
42	0.563	0.604	0.670	0.733	0.813	0.875	0.930	0.984
44	0.572	0.612	0.677	0.738	0.818	0.878	0.932	0.985
46	0.580	0.620	0.683	0.744	0.822	0.881	0.934	0.986
48	0.587	0.627	0.690	0.749	0.825	0.884	0.936	0.986
50	0.594	0.633	0.695	0.754	0.829	0.886	0.937	0.987
60	0.625	0.662	0.720	0.774	0.844	0.897	0.944	0.989
70	0.649	0.684	0.739	0.790	0.856	0.905	0.949	0.990
80	0.669	0.703	0.755	0.803	0.865	0.911	0.952	0.992
90	0.686	0.718	0.768	0.814	0.873	0.917	0.955	0.993
100	0.701	0.731	0.779	0.824	0.879	0.921	0.958	0.993
120	0.724	0.753	0.798	0.839	0.890	0.928	0.962	0.994
140	0.743	0.770	0.812	0.850	0.898	0.934	0.965	0.995
160	0.758	0.784	0.823	0.860	0.905	0.938	0.968	0.996
180	0.771	0.796	0.833	0.868	0.910	0.942	0.970	0.996
200	0.782	0.806	0.841	0.874	0.915	0.945	0.972	0.997

Table C-4 χ^2 distribution (*continued*)

P ν	0.40	0.30	0.20	0.10	0.05	0.02	0.01	0.001
1	0.708	1.074	1.642	2.706	3.841	5.412	6.635	10.827
2	0.916	1.204	1.609	2.303	2.996	3.912	4.605	6.908
3	0.982	1.222	1.547	2.084	2.605	3.279	3.780	5.423
4	1.011	1.220	1.497	1.945	2.372	2.917	3.319	4.617
5	1.026	1.213	1.458	1.847	2.214	2.678	3.017	4.102
6	1.035	1.205	1.426	1.774	2.099	2.506	2.802	3.743
7	1.040	1.198	1.400	1.717	2.010	2.375	2.639	3.475
8	1.044	1.191	1.379	1.670	1.938	2.271	2.511	3.266
9	1.046	1.184	1.360	1.632	1.880	2.187	2.407	3.097
10	1.047	1.178	1.344	1.599	1.831	2.116	2.321	2.959
11	1.048	1.173	1.330	1.570	1.789	2.056	2.248	2.842
12	1.049	1.168	1.318	1.546	1.752	2.004	2.185	2.742
13	1.049	1.163	1.307	1.524	1.720	1.959	2.130	2.656
14	1.049	1.159	1.296	1.505	1.692	1.919	2.082	2.580
15	1.049	1.155	1.287	1.487	1.666	1.884	2.039	2.513
16	1.049	1.151	1.279	1.471	1.644	1.852	2.000	2.453
17	1.048	1.148	1.271	1.457	1.623	1.823	1.965	2.399
18	1.048	1.145	1.264	1.444	1.604	1.797	1.934	2.351
19	1.048	1.142	1.258	1.432	1.586	1.773	1.905	2.307
20	1.048	1.139	1.252	1.421	1.571	1.751	1.878	2.266
22	1.047	1.134	1.241	1.401	1.542	1.712	1.831	2.194
24	1.046	1.129	1.231	1.383	1.517	1.678	1.791	2.132
26	1.045	1.125	1.223	1.368	1.496	1.648	1.755	2.079
28	1.045	1.121	1.215	1.354	1.476	1.622	1.724	2.032
30	1.044	1.118	1.208	1.342	1.459	1.599	1.696	1.990
32	1.043	1.115	1.202	1.331	1.444	1.578	1.671	1.953
34	1.042	1.112	1.196	1.321	1.429	1.559	1.649	1.919
36	1.042	1.109	1.191	1.311	1.417	1.541	1.628	1.888
38	1.041	1.106	1.186	1.303	1.405	1.525	1.610	1.861
40	1.041	1.104	1.182	1.295	1.394	1.511	1.592	1.835
42	1.040	1.102	1.178	1.288	1.384	1.497	1.576	1.812
44	1.039	1.100	1.174	1.281	1.375	1.485	1.562	1.790
46	1.039	1.098	1.170	1.275	1.366	1.473	1.548	1.770
48	1.038	1.096	1.167	1.269	1.358	1.462	1.535	1.751
50	1.038	1.094	1.163	1.263	1.350	1.452	1.523	1.733
60	1.036	1.087	1.150	1.240	1.318	1.410	1.473	1.660
70	1.034	1.081	1.139	1.222	1.293	1.377	1.435	1.605
80	1.032	1.076	1.130	1.207	1.273	1.351	1.404	1.560
90	1.031	1.072	1.123	1.195	1.257	1.329	1.379	1.525
100	1.029	1.069	1.117	1.185	1.243	1.311	1.358	1.494
120	1.027	1.063	1.107	1.169	1.221	1.283	1.325	1.446
140	1.026	1.059	1.099	1.156	1.204	1.261	1.299	1.410
160	1.024	1.055	1.093	1.146	1.191	1.243	1.278	1.381
180	1.023	1.052	1.087	1.137	1.179	1.228	1.261	1.358
200	1.022	1.050	1.083	1.130	1.170	1.216	1.247	1.338

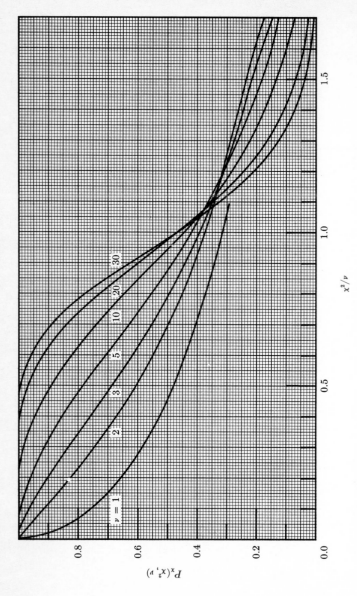

FIGURE C-4 The probability $P_\chi(\chi^2, \nu)$ of exceeding χ^2 vs. the reduced chi-square $\chi^2_\nu = \chi^2/\nu$ and the number of degrees of freedom ν.

C-5 F DISTRIBUTION

The probability distribution for F is given by

$$P_f(f, \nu_1, \nu_2) = \frac{\Gamma[(\nu_1 + \nu_2)/2]}{\Gamma(\nu_1/2)\Gamma(\nu_2/2)} \left(\frac{\nu_1}{\nu_2}\right)^{1/2} \frac{f^{\frac{1}{2}(\nu_1-2)}}{(1 + f\nu_1/\nu_2)^{\frac{1}{2}(\nu_1 + \nu_2)}}$$

The probability of observing a value of F-test larger than F for a random sample with ν_1 and ν_2 degrees of freedom is the integral of this probability.

$$P_F(F, \nu_1, \nu_2) = \int_F^\infty P_f(f, \nu_1, \nu_2) \, df$$

Values of F corresponding to various values of the integral probability $P_F(F, \nu_1, \nu_2)$ of exceeding F in a measurement are tabulated in Table C-5 for $\nu_1 = 1$ and graphed in Figure C-5 as a smooth variation with the probability. Values of F corresponding to various values of ν_1 and ν_2 ranging from 1 to ∞ are listed in Table C-6 and graphed in Figure C-6 for $P_F(F, \nu_1, \nu_2) = 0.05$ and in Table C-7 and Figure C-7 for $P_F(F, \nu_1, \nu_2) = 0.01$. These values were adapted by permission from Dixon and Massey.

The function which is tabulated and graphed is the shaded area under the tail of the probability curve for values larger than F as indicated.

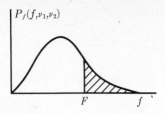

$P_f(f,\nu_1,\nu_2)$

F f

Table C-5 F distribution, $\nu = 1$. Values of F corresponding to the probability $P_F(F,1,\nu_2)$ of exceeding F (with $\nu_1 = 1$ degrees of freedom) vs. the larger number of degrees of freedom ν_2

Degrees of freedom ν_2	Probability of exceeding F							
	$P = 0.50$	0.25	0.10	0.05	0.025	0.01	0.005	0.001
1	1.000	5.83	39.9	161	648	4050	16200	406000
2	.667	2.57	8.53	18.5	38.5	98.5	198	998
3	.585	2.02	5.54	10.1	17.4	34.1	55.6	167
4	.549	1.81	4.54	7.71	12.2	21.2	31.3	74.1
5	.528	1.69	4.06	6.61	10.0	16.3	22.8	47.2
6	.515	1.62	3.78	5.99	8.81	13.7	18.6	35.5
7	.506	1.57	3.59	5.59	8.07	12.2	16.2	29.2
8	.499	1.54	3.46	5.32	7.57	11.3	14.7	25.4
9	.494	1.51	3.36	5.12	7.21	10.6	13.6	22.9
10	.490	1.49	3.28	4.96	6.94	10.0	12.8	21.0
11	.486	1.47	3.23	4.84	6.72	9.65	12.2	19.7
12	.484	1.46	3.18	4.75	6.55	9.33	11.8	18.6
15	.478	1.43	3.07	4.54	6.20	8.68	10.8	16.6
20	.472	1.40	2.97	4.35	5.87	8.10	9.94	14.8
24	.469	1.39	2.93	4.26	5.72	7.82	9.55	14.0
30	.466	1.38	2.88	4.17	5.57	7.56	9.18	13.3
40	.463	1.36	2.84	4.08	5.42	7.31	8.83	12.6
60	.461	1.35	2.79	4.00	5.29	7.08	8.49	12.0
120	.458	1.34	2.75	3.92	5.15	6.85	8.18	11.4
∞	.455	1.32	2.71	3.84	5.02	6.63	7.88	10.8

For larger values of the probability P, the value of F is approximately $F \simeq [1.25(1 - P)]^2$.

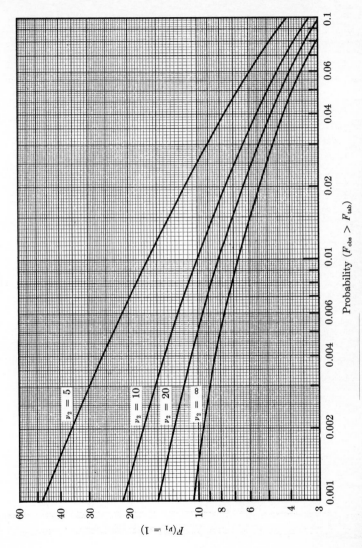

FIGURE C-5 The probability $P_F(F, 1, \nu_2)$ of exceeding F vs. F and ν_2 for $\nu_1 = 1$.

Table C-6 F distribution, 5%. Values of F corresponding to the probability $P_F(F, \nu_1, \nu_2) = 0.05$ of exceeding F for ν_1 vs. ν_2 degrees of freedom

Degrees of freedom ν_2	Degrees of freedom ν_1							
	2	4	6	8	10	15	20	100
1	200	225	234	239	242	246	248	253
2	19.0	19.2	19.3	19.4	19.4	19.4	19.4	19.5
3	9.55	9.12	8.94	8.85	8.79	8.70	8.66	8.55
4	6.94	6.39	6.16	6.04	5.96	5.86	5.80	5.66
5	5.79	5.19	4.95	4.82	4.73	4.62	4.56	4.41
6	5.14	4.53	4.28	4.15	4.60	3.94	3.87	3.71
7	4.74	4.12	3.87	3.73	3.64	3.51	3.44	3.27
8	4.46	3.84	3.58	3.44	3.35	3.22	3.15	2.97
9	4.26	3.63	3.37	3.23	3.14	3.01	2.94	2.76
10	4.10	3.48	3.22	3.07	2.98	2.85	2.77	2.59
11	3.98	3.36	3.09	2.95	2.85	2.72	2.65	2.46
12	3.89	3.26	3.00	2.85	2.75	2.62	2.54	2.35
15	3.68	3.06	2.79	2.64	2.54	2.40	2.33	2.12
20	3.49	2.87	2.60	2.45	2.35	2.20	2.12	1.91
24	3.40	2.78	2.51	2.36	2.25	2.11	2.03	1.80
30	3.32	2.69	2.42	2.27	2.16	2.01	1.93	1.70
40	3.23	2.61	2.34	2.18	1.08	1.92	1.84	1.59
60	3.15	2.53	2.25	2.10	1.99	1.84	1.75	1.48
120	3.07	2.45	2.18	2.02	1.91	1.75	1.66	1.37
∞	3.00	2.37	2.10	1.94	1.83	1.67	1.57	1.24

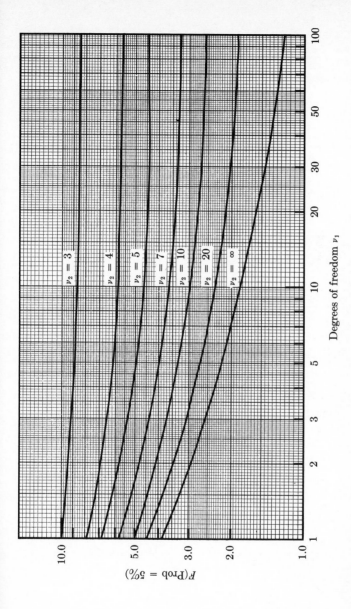

FIGURE C-6 Test values of $F(\nu_1, \nu_2)$ vs. the numbers of degrees of freedom ν_1 and ν_2 for a probability $P_F(F, \nu_1, \nu_2) = 0.05$ of exceeding F.

Table C-7 F distribution, 1%. Values of F corresponding to the probability $P_F(F, \nu_1, \nu_2) = 0.01$ of exceeding F for ν_1 vs. ν_2 degrees of freedom

Degrees of freedom ν_2	Degrees of freedom ν_1							
	2	4	6	8	10	15	20	100
1	5000	5620	5860	5980	6060	6160	6210	6330
2	99.0	99.2	99.3	99.4	99.4	99.4	99.4	99.5
3	30.8	28.7	27.9	27.5	27.2	26.9	26.7	26.2
4	18.0	16.0	15.2	14.8	14.5	14.2	14.0	13.6
5	13.3	11.4	10.7	10.3	10.1	9.72	9.55	9.13
6	10.9	9.15	8.47	8.10	7.87	7.56	7.40	6.99
7	9.55	7.85	7.19	6.84	6.62	6.31	6.16	5.75
8	8.65	7.01	6.37	6.03	5.81	5.52	5.36	4.96
9	8.02	6.42	5.80	5.47	5.26	4.96	4.81	4.42
10	7.56	5.99	5.39	5.06	4.85	4.56	4.41	4.01
11	7.21	5.67	5.07	4.74	4.54	4.25	4.10	3.71
12	6.93	5.41	4.82	4.50	4.30	4.01	3.86	3.47
15	6.36	4.89	4.32	4.00	3.80	3.52	3.37	2.98
20	5.85	4.43	3.87	3:56	3.37	3.09	2.94	2.54
24	5.61	4.22	3.67	3.36	3.17	2.89	2.74	2.33
30	5.39	4.02	3.47	3.17	2.98	2.70	2.55	2.13
40	5.18	3.83	3.29	2.99	2.80	2.52	2.37	1.94
60	4.98	3.65	3.12	2.82	2.63	2.35	2.20	1.75
120	4.79	3.48	2.96	2.66	2.47	2.19	2.03	1.56
∞	4.61	3.32	2.80	2.51	2.32	2.04	1.88	1.36

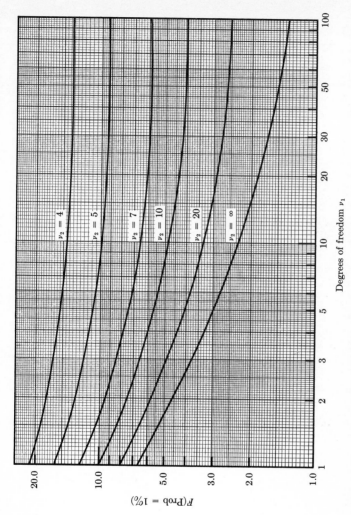

FIGURE C-7 Test values of $F(\nu_1, \nu_2)$ vs. the numbers of degrees of freedom ν_1 and ν_2 for a probability $P_F(F, \nu_1, \nu_2) = 0.01$ of exceeding F.

REFERENCES

Anderson, R. L., and E. E. Houseman: Tables of Orthogonal Polynomial Values Extended to N = 104, *Res. Bull.* 297, *Agr. Exp. Sta.*, *Iowa State Univ.*, April, 1942.

Arndt, Richard A., and Malcolm H. MacGregor: Nucleon-Nucleon Phase Shift Analyses by Chi-Squared Minimization, "Methods in Computational Physics," vol. 6, pp. 253–296, Academic Press Inc., New York, 1966.

Baird, D. C.: "Experimentation: An Introduction to Measurement Theory and Experiment Design," Prentice-Hall, Inc., Englewood Cliffs, N.J., 1962.

Beers, Yardley: "Introduction to the Theory of Error," 2d ed., Addison-Wesley Publishing Company, Inc., Reading, Mass., 1957.

Colcord, C. C., and L. S. Deming: The One-Tenth Percent Level of Z, *Sankhya*, vol. 2, pt. 4, pp. 423–424, December, 1936.

Cziffra, Peter, and Michael J. Moravscik: A Practical Guide to the Method of Least Squares, *UCRL*-8523, University of California Radiation Laboratory, Berkeley, Calif., 1958.

David, F. N.: "Tables of the Correlation Coefficients," Cambridge University Press, London, 1938.

Dixon, W. J., and F. J. Massey, Jr.: "Introduction to Statistical Analysis," 3d ed., McGraw-Hill Book Company, New York, 1969.

Evans, R. E.: "The Atomic Nucleus," chaps. 26–28 and appendix G, McGraw-Hill Book Company, New York, 1962.

Hald, A., and S. A. Sinkbaek: A Table of Percentage Points of the χ^2-Distribution, *Skand. Aktuarietidskrift*, pp. 170–175, 1950.

Hamilton, Walter Clark: "Statistics in Physical Science," The Ronald Press Company, New York, 1964.

"Handbook of Chemistry and Physics," Chemical Rubber Publishing Co., Cleveland, Ohio, 1968.

Hoel, Paul G.: "Introduction to Mathematical Statistics," 2d ed., John Wiley & Sons, Inc., New York, 1954.

IBM, "System/360 Scientific Subroutine Package" (360A-CM-03X), Programmer's Manual.

Marquardt, Donald W.: An Algorithm for Least-Squares Estimation of Nonlinear Parameters, *J. Soc. Ind. Appl. Math.*, vol. 11, no. 2, pp. 431–441, June, 1963.

Melkanoff, Michel A., Tatsuro Sawada, and Jacques Raynal: Nuclear Optical Model Calculations, *Methods in Computational Physics*, vol. 6, pp. 2–80, Academic Press Inc., New York, 1966.

Merrington, M., and C. M. Thompson: Tables of Percentage Points of the Inverted Beta (F) Distribution, *Biometrika*, vol. 33, pt. 1, pp. 74–87, April, 1943.

Orear, Jay: "Notes on Statistics for Physicists," *UCRL*-8417, University of California Radiation Laboratory, Berkeley, Calif., 1958.

Ostle, Bernard: "Statistics in Research," 2d ed., Iowa State College Press, Ames, Iowa, 1963.

Pearson, Karl: "Tables for Statisticians and Biometricians," Cambridge University Press, London, 1924.

Pugh, Emerson M., and George H. Winslow: "The Analysis of Physical Measurements," Addison-Wesley Publishing Company, Inc., Reading, Mass., 1966.

Snedecor, G. W.: "Statistical Methods," 4th ed., Iowa State College Press, Ames, Iowa, 1946.

Thrall, Robert M., and Leonard Tornheim: "Vector Spaces and Matrices," John Wiley & Sons, Inc., New York, 1957.

Tyapkin, A. A.: "Phase Shift Analysis of p-p Scattering at 95, 150, and 310 MeV," Proceedings of the 1960 Annual International Conference on High Energy Physics at Rochester, pp. 138–140, Interscience Publishers, Inc., New York, 1960.

Wilf, Herbert S.: "Calculus and Linear Algebra," Harcourt, Brace & World, Inc., New York, 1966.

Worthing, Archie, and Joseph Gaffner: "Treatment of Experimental Data," John Wiley & Sons, Inc., New York, 1943.

Young, Hugh D.: "Statistical Treatment of Experimental Data," McGraw-Hill Book Company, New York, 1962.

SOLUTIONS TO
ODD-NUMBERED EXERCISES

CHAPTER 1

1-1 (*a*) 5   (*b*) 2   (*c*) 2   (*d*) 5   (*e*) 4
     (*f*) 1   (*g*) 3   (*h*) 3   (*i*) 3   (*j*) 4
1-3 (*a*) 4.5   (*b*) 1.8   (*c*) 2.0   (*d*) 1.0   (*e*) 1.8
     (*f*) 10.   (*g*) 7.4   (*h*) 1.0×10^2 (*i*) 14   (*j*) 4.7

CHAPTER 2

2-1 Mean = 7.65; median = 8; most probable value = 8
2-3 $s = 2.28$
2-5 Mean = 7; median = 7; standard deviation = 2.42
2-7 $C = 4/R^3$

2-9 $\dfrac{1}{N} \Sigma(x_i - \mu)^2 = \dfrac{1}{N} \Sigma x_i^2 - \dfrac{2}{N} \mu \Sigma x_i + \mu^2$

$$= \dfrac{1}{N} \Sigma x_i^2 - 2\mu^2 + \mu^2$$

2-11 $\displaystyle\sum_{j=1}^{n} [(x_j - \mu)^2 P(x_j)] = \sum_{j=1}^{n} [x_j^2 P(x_j)] - 2\mu \sum_{j=1}^{n} [x_j P(x_j)] - \mu^2$

$$= \sum_{j=1}^{n} [x_j^2 P(x_j)] - 2\mu^2 + \mu^2$$

CHAPTER 3

3-1 (a) 6   (b) 120   (c) 6720   (d) 462   (e) 3,628,800

3-3 (a) 20   (b) 6   (c) 120   (d) 270,725

3-5 41 cents for one lemon; \$3.70 for two lemons; \$100 for three lemons

3-7 $P_P(\mu,\mu) = \dfrac{\mu^\mu}{\mu!} e^{-\mu} = \dfrac{\mu(\mu^{\mu-1})}{\mu(\mu-1)!} e^{-\mu} = \dfrac{\mu^{\mu-1}}{(\mu-1)!} e^{-\mu} = P_P(\mu - 1, \mu)$

3-9 $P_G(\mu + \sigma, \mu, \sigma) = 0.24197 = 0.60653 P_G(\mu,\mu,\sigma)$

$P_G(\mu + \text{P.E.}, \mu, \sigma) = 0.3178 = 0.7965 P_G(\mu,\mu,\sigma)$

$P_G(\mu + \Gamma/2, \mu, \sigma) = 0.19947 = \tfrac{1}{2} P_G(\mu,\mu,\sigma)$

3-11 Area $= \dfrac{2}{\pi} \tan^{-1}(3) = 0.795$

3-13 $\sigma^2 = \displaystyle\sum_{x=0}^{n} \left[x^2 \dfrac{n!}{x!(n-x)!} p^x (1-p)^{n-x} \right] - \mu^2$

$$= \sum_{x=0}^{n} \left[x \dfrac{n!}{(x-1)!(n-x)!} p^x (1-p)^{n-x} \right] - \mu^2$$

$$= \sum_{y=0}^{n} \left[(y+1) \dfrac{(n)m!}{y!(m-y)!} p^{y+1} (1-p)^{m-y} \right] - \mu^2$$

$$= np(mp+1) - \mu^2 = np(np - p + 1) - (np)^2 = np(1-p)$$

CHAPTER 4

4-1 (a) $\sigma_x^2 = \tfrac{1}{4}(\sigma_u^2 + \sigma_v^2)$ (b) $\sigma_x^2 = \tfrac{1}{4}(\sigma_u^2 + \sigma_v^2)$

(c) $\sigma_x = 2\sigma_u/u^3$ (d) $\sigma_x^2 = v^4\sigma_u^2 + 4u^2v^2\sigma_v^2$

(e) $\sigma_x^2 = 4u^2\sigma_u^2 + 4v^2\sigma_v^2$

4-3 For equal contributions to the total uncertainty, $\sigma_r/r = \tfrac{1}{2}\sigma_L/L$

CHAPTER 5

5-1 $\sigma_u \simeq s/\sqrt{N} = 0.048$ cm; P.E. $= 0.6745\sigma_\mu = 0.032$ cm
$= \frac{3}{4}(\bar{x} - \mu);$ $A_G(\bar{x},\mu,\sigma_\mu) = 0.439 = 56\%$ probable to have so
large a discrepancy

5-3 $\sigma \simeq s = 12.5;$ $\sigma_\mu \simeq s/\sqrt{N} = 1.97$

5-5 $\bar{x} = 1.95;$ $s = 0.30;$ $s_\mu = 0.10$

5-7 $R = (910 - 780)/5 = 166$ counts/min; $\sigma_R \simeq \sqrt{910 + 78}/5 = 6.2$ counts/min

5-9 $\chi^2 \simeq 8(0.93 \pm 0.3) = 7.5 \pm 2.4;$ $\nu = 8$

CHAPTER 6

6-1 $a = 104.4,$ $b = -0.58/\text{sec};$ $s = 8.7,$ $\sigma_a \simeq 5.1,$ $\sigma_b \simeq 0.064/\text{sec}$

6-3 $a = 4.17°\text{C},$ $b = 9.40°\text{C/cm};$ $\sigma_a \simeq 3.5°\text{C},$ $\sigma_b \simeq 0.83°\text{C/cm}$

6-5 The average deviation is only $\frac{3}{4}s$. The largest deviation is $1.43s$, which has a probability of occurrence of $\frac{1}{7}$. Since there are 10 data points, such a discrepancy is quite likely.

6-7 $b = [\Sigma(x_i y_i/\sigma_i^2)]/\Sigma(x_i^2/\sigma_i^2)$

CHAPTER 7

7-1 $r = 0.97$

7-3 $r = 0.97$

7-5 $r_1 = 0.9992$

7-7 $r_2 = 0.9749$

7-9 $R = 0.99996$

CHAPTER 8

8-1 Thirteen of the 21 discrepancies are less than or equal to s. The largest discrepancy is $2.5s$, which has a probability of occurrence of $\frac{1}{80}$. What is surprising is that there is another discrepancy (with the opposite sign) which is almost as large.

8-3 $C = (879 \pm 5) - (423 \pm 80)\cos^2\theta + (901 \pm 81)\cos^4\theta$

8-5 $P_4(x) = \frac{1}{8}(35x^4 - 30x^2 + 3)$

8-7 $C = 1. + (0.254 \pm .020)P_2 + (0.224 \pm .020)P_4$

CHAPTER 9

9-1 Yes, the methods are essentially equivalent.

9-3 It would be reasonable for the values of the coefficients b_j to fall within an order of magnitude of 1, but the actual range cannot be predicted from the range of r_{jy} and r_{jk}. The coefficients b_j can have any finite value.

9-5 $1/y = a + bx$; $\sigma_i' = -\sigma_i/y_i^2$

CHAPTER 10

10-1 Approximately 10% probability

10-3 Approximately 0.1% probability; presumably not a good fit

10-5 $F \simeq 10$ for $\nu_1 = 1$; $F \simeq 5$ for $\nu_1 \lesssim \nu_2$

CHAPTER 13

13-1
$n =$	1	2	3	4	5	6
$y =$	823	869	856	841	805	767

Legendre polynomial $y = 829$

Note that interpolation is over cosines, not degrees.

13-2
$n =$	1	2	3	4	5	6
$A =$	1811	1761	1763	1897	2329	3147

Legendre polynomial area $= 1836$

Note that integration is for $-1 \leq \cos \theta \leq 1$.

INDEX

Boldface page numbers refer to summaries at the ends of chapters.